INSTRUCTOR'S GUIDE/TEST BANK FOR

Essentials of Human Anatomy and Physiology

Fifth Edition

Elaine N. Marieb, R. N., Ph.D.

with contributions from

Patricia J. Turner
Howard Community College

An Imprint of Addison Wesley Longman, Inc.

Menlo Park, California • Reading, Massachusetts • New York • Harlow, England
Don Mills, Ontario • Sydney • Mexico City • Madrid • Amsterdam

Assistant Editor: Thor Ekstrom
Production Coordinator: David Novak/Wendy Earl Productions
Proofreader: Martha Ghent
Cover Designer: Vargas/Williams Design

ISBN 0-8053-4072-6

4 5 6 7 8 9 10–VG–00 99

The Benjamin/Cummings Publishing Company, Inc.
2725 Sand Hill Road
Menlo Park, California 94025

CONTENTS

PREFACE

The Instructor's Guide to accompany the Fifth Edition of *Essentials of Human Anatomy and Physiology* by Elaine N. Marieb features a wealth of information for the anatomy and physiology instructor.

Organization of this Guide

This guide is divided into two sections. The first section, Instructional Aids, includes five subsections for each chapter: Suggested Lecture Outlines, Lecture Hints, Answers to Textbook Review Questions, Classroom Demonstrations and Student Activities, and Audiovisual Aids. The second section, the Test Bank, contains sample test questions for each chapter.

Suggested Lecture Outlines. A new feature of this edition of the *Instructor's Guide*, these outlines provide a quick overview of the chapter contents.

Lecture Hints. Also new to this edition, the lecture hints have useful information such as topics that are common student pitfalls, creative tips for clarifying difficult concepts with analogies and short demonstrations, and ideas for discussion topics that can add interest to the class.

Answers to Textbook Review Questions. To assist students in studying, review questions appear at the end of each textbook chapter. This section also includes answers to the "At the Clinic" questions.

Classroom Demonstrations and Student Activities. The suggestions provided here address the need to have students actively involved in their learning. The activities run the gamut from participatory group activities to library research topics. Because most courses for this text do not have lab sessions, many of the demonstrations included here use models and simple laboratory equipment. Hopefully, these student-oriented activities will enable them to solidify their understanding or substantially expand it.

Audiovisual Aids. This section contains listings of films, videodiscs, and computer software. Each title description includes the following information, when available, in the sequence indicated: title, distributor code, length of title, and year of release. If a listing is of particular interest, it is recommended that you check your own college's catalog of audiovisual materials. If the title is not available there, request it from the distributor listed. The distributors' addresses and phone numbers are listed in the Audiovisual Resources Guide that appears at the end of the Instructional Aids section of this guide.

Test Bank. The test bank contains questions in multiple-choice, matching, and true/false formats. This test bank is intended to provide a foundation of test items of varying difficulty that can be used as written or modified to correlate with the personal focus of your own class.

Special thanks to Patricia Turner of Howard Community College, Columbia, Maryland, for providing the lecture hints, supplementing the classroom demonstrations and student activities, and updating the audiovisual aids. Her contributions are an excellent addition to this guide, and her hard work is much appreciated.

Part I   Instructor's Guide

1 The Human Body: An Orientation

SUGGESTED LECTURE OUTLINE

   I. AN OVERVIEW OF ANATOMY AND PHYSIOLOGY (p. 2)

      A. Anatomy (p. 2)

      B. Physiology (p. 2)

      C. Relationship Between Anatomy and Physiology (p. 2)

   II. LEVELS OF STRUCTURAL ORGANIZATION (pp. 2–7)

      A. From Atoms to Organisms (pp. 2–4)

      B. Organ System Overview (pp. 4–7)

         1. Integumentary System

         2. Skeletal System

         3. Muscular System

         4. Nervous System

         5. Endocrine System

         6. Circulatory System

         7. Respiratory System

         8. Digestive System

         9. Urinary System

       10. Reproductive System

  III. MAINTAINING LIFE (pp. 7–9)

      A. Necessary Life Functions (pp. 7–9)

         1. Maintaining Boundaries

         2. Movement

         3. Responsiveness

         4. Digestion

         5. Metabolism

         6. Excretion

         7. Reproduction

         8. Growth

      B. Survival Needs (p. 9)

  IV. HOMEOSTASIS (pp. 9–10)

      A. Homeostatic Control Mechanisms (p. 10)

      B. Homeostatic Imbalance (p. 10)

   V. THE LANGUAGE OF ANATOMY (pp. 11–17)

      A. Anatomical Position and Directional Terms (pp. 11–12)

      B. Regional Terms (pp. 11–13)

         1. Anterior Body Landmarks

         2. Posterior Body Landmarks

      C. Body Planes and Sections (pp. 13–15)

      D. Body Cavities (pp. 15–17)

         1. Dorsal Body Cavity

         2. Ventral Body Cavity

LECTURE HINTS

1. Explore the meaning of the terms anatomy and physiology and discuss how they are typically studied by scientists.

2. Emphasize that part of this course will involve learning a vast new language. Instruct students to make pocket-sized flash cards for any term they encounter in their reading whose meaning is not immediately known. Students can then carry the flash cards around and review them frequently during the day. Any term can be discarded from the pile when it is mastered.

3. Emphasize that the language of anatomy is often redundant and that multiple terms exist for structures or events. These terms come from tradition (national origin, discoverer) or were named in a more practical fashion, based on structure or function. Often both structural and functional names are given. Make sure students realize that it is difficult to master all of these terms quickly and can be frustrating as well, but that they will learn them completely with experience.

4. Provide students with opportunities to verbalize material you have covered in your lecture. The unfamiliar terms students are introduced to will have more meaning if they say them out loud. Encourage students to study "out loud," practicing vocabulary and explaining concepts. Encourage students to explain processes or mechanisms in their own words whenever possible.

5. Teach students the appropriate word roots whenever possible. In addition to helping students figure out the meaning of new words, these roots are often based on very interesting ideas.

6. Point out to students how the terms anterior/posterior and superior/inferior refer to different areas for bipeds and quadrupeds. Also point out that more than one word can be used to correctly describe the position of a single structure.

7. Most people correctly visualize a space when they hear the term cavity. Most body cavities, however, are potential spaces and are filled with viscera, tissues, and fluids. In addition to the dorsal body cavity and the ventral body cavity, mention other cavities such as the oral cavity, the nasal cavity, and the pericardial cavity.

ANSWERS TO TEXTBOOK REVIEW QUESTIONS
(Questions appear on pp. 19–20)

1. Anatomy: Study of the structure and shape of body parts and their relationship to one another.

   Physiology: Study of the function of the body or body parts, that is, how they work or operate.

2. The structure or anatomy of body parts and organs determines what they are able to do and how they can function. If you do not understand the structure of a body part, it is extremely difficult to understand its operation.

3. Atom, cell, tissue, organ, organ system, organism.

4. a. Integumentary system: Functions basically to protect. The integumentary system is an organ and an organ system. (Students may name sweat glands and the like, but this is not really necessary.)

   b. Skeletal system: Major role is to support the body and provide a framework on which the muscles can act to create movement; also protects by enclosing. Bones, ligaments.

c. Muscular system: Major role is to promote body movement. Organs include the skeletal muscles.

d. Nervous system: The body's fast-acting controlling and coordinating system; acts via electrical signals called nerve impulses. Brain, spinal cord, nerves.

e. Endocrine system: The body's slower-acting controlling and coordinating system, which acts through chemicals called hormones. Pancreas, adrenals, thyroid, pituitary and others.

f. Cardiovascular system: Basically a transport and delivery system for bringing to the body's cells adequate supplies of oxygen and nutrients and for getting rid of cell wastes. Heart, blood vessels.

g. Lymphatic system: Collects fluid leaked from the cardiovascular system and returns it to the bloodstream. Houses cells involved in the immune response. Lymphatic vessels, lymph nodes. Together with the cardiovascular system, comprises the circulatory system.

h. Respiratory system: Exchanges respiratory gases, that is, takes in oxygen and releases carbon dioxide to the body exterior. Lungs, bronchi, bronchioles, trachea.

i. Digestive system: Breaks down ingested foods so that they can be absorbed into the bloodstream and thereby made available to the body's cells. Stomach, small intestine, large intestine.

j. Urinary system: Rids the body of nitrogenous wastes and regulates the acid-base and water balance of the body fluids. Kidneys, ureters, urethra.

k. Reproductive system: Produces sex (germ) cells so that reproduction of the individual can occur. Ovaries or testes, plus various duct system structures.

5. Other functional requirements are the ability to move, to maintain boundaries, to reproduce, and to respond to stimuli.

6. Food, oxygen, water, appropriate temperature, and normal atmospheric pressure.

7. Homeostasis: A relatively stable condition of the body's internal environment that is maintained by various functional mechanisms.

8. Aging, abnormal body functioning leading to illness and disease and/or death.

9. Anatomical position: Erect, arms at sides, with palms and toes facing anteriorly.

10. Nose—anterior surface; calf—posterior surface; umbilicus—anterior surface; fingernails—posterior surface.

11. a. Foot; b. Cheekbone; c. Chin; d. Toenails; e. Skin.

12. Frontal (coronal).

13. Both subdivisions of the ventral body cavity—circulatory, muscular (diaphragm), and digestive. Thoracic cavity only—respiratory. Abdominopelvic cavity only—reproductive and urinary.

14. See Figure 1.7b, p. 17 of the text.

ANSWERS TO AT THE CLINIC QUESTIONS

1. Antecubital region: He should have held out the anterior surface of his arm. (The antecubital region is the anterior aspect of the elbow.)

   Deltoid region: He should have taken off his shirt to receive a shot in the shoulder region.

   Sural: His left calf was bruised.

2. a. With age, body organs and control systems become less efficient. The drop in efficiency causes the internal environment to be less and less stable.

   b. Examples include: bone loss with age, possibly due to a lack of continued stimulation of osteoblasts by hormones; senescent cells contain a chemical that may be responsible for cellular aging; decrease in muscle fibers and increase in connective tissue of the skeletal muscles; and sclerosis and thickening of heart valve flaps possibly due to stress of blood flow.

3. When we take a drink, body hydration increases and thirst declines—an example of a typical negative feedback system. If it were a positive feedback system, the body's need for water (and thirst) would increase after we take a drink.

4. Jane's nerve is injured where her upper appendage attaches to the trunk (armpit). Torn ligaments are located in her neck and shoulder blade region. The broken bone is located in the right upper appendage between the shoulder and elbow.

CLASSROOM DEMONSTRATIONS AND STUDENT ACTIVITIES

Classroom Demonstrations

1. Film(s) or other AV materials of choice.

2. Dissect a freshly killed rat or a small preserved animal (e.g., a fetal pig) to demonstrate the various organ systems of the body. Point out at least two organs from each organ system and discuss their function. If dissection is not an option, there are several films that can be substituted (see Audiovisual Aids section). In addition, Carolina Biological Supply Company offers preserved dissected specimens that are mounted in clear acrylic containers.

3. Use a dissectible human torso model to point out the dorsal and ventral body cavities and the organs in each cavity.

4. Arrange for the class to attend an autopsy (preferably after the material of Chapter 1 has been covered).

5. Decorate your classroom with posters and diagrams of body systems. Label with appropriate directional terms, regional terms, or terms related to body planes or body cavities.

6. Thin, plastinated sections of the human body can be purchased from Carolina Biological Supply Company and can be viewed with the naked eye, or projected on an overhead projector for viewing by the entire class.

Student Activities

1. Assume the anatomical position. Ask the students to comment on how that position differs from the "usual" standing position and to explain why knowing this position is important to them in their current studies.

2. Remove all the organs from the ventral and dorsal body cavities of a human torso model. Ask for volunteers or assign students to return them to their proper anatomic location. As each organ is properly repositioned, the rest of the students are to call out its name and organ system relationship.

3. Place a chair at center stage. Ask for a volunteer to come up and show how the chair would be cut along the sagittal, frontal, and transverse planes and to choose which of these planes would yield a "usable" seat, providing an explanation for the choice made. (The explanation should also include the reasons the other options were not selected.)

4. To initiate a class discussion on the relative degree of protection of organs in the dorsal versus the ventral body cavity, ask the class a question such as, "Why do you think a dog instinctively curls over and protects its abdomen when that body region is approached (threatened by a blow, etc.) even playfully?" or "Two people have rapidly growing tumors. The tumor is in the dorsal cavity in one of these individuals and in the ventral cavity of the other. Which of these people will develop symptoms first and why?"

5. Call out anatomical terms (buccal, femoral, etc.) and have the students (as a group) point out the named regions on their own bodies.

6. To help the students understand negative feedback, ask them to explain how scratching an itch can be considered an example of negative feedback.

7. To encourage an understanding of organ system interrelationships, ask the students to comment on the functional relationships between muscles and bones, and between the respiratory and circulatory systems.

8. Use a simple battery-operated clock as an example to indicate the importance of relatively constant conditions on optimal functioning of any system. Ask the class to indicate what would happen if it was (a) immersed in water, (b) hit with a hammer, (c) heated until it was red-hot, and (d) had its battery removed. Then ask them what conditions would be best suited for the clock to operate properly, and ask them to provide a single term that describes those conditions (homeostasis).

9. Explore with the class the meaning of "alive." Use examples of animals exhibiting these characteristics on film or videodisc, and ask students to identify the characteristics being demonstrated. Or, allow students to develop a list of characteristics by discussing their own life needs.

10. Student assignment for class discussion: Bring in an article from a popular magazine (*Time, Newsweek*, etc.) describing an environmental problem(s), such as toxic waste disposal, pollution of the ocean, etc., that threatens their homeostasis, even survival. Be prepared to describe the problem and how it represents a threat to the body.

11. If you have a magnetic board in your classroom, you can make signs listing various organs of the body and glue magnets to them. Have students group the organs into appropriate body systems. These magnets can be purchased inexpensively from hardware or stationery supply stores.

AUDIOVISUAL AIDS

Videotapes

1. *Anatomy of the Fetal Pig* (FSH, VHS, 62 min.)
2. *Dissection and Anatomy of the Cat* (ALT, 1988)
3. *Dissection and Anatomy of the Fetal Pig* (ALT, 1988)
4. *Homeostasis* (FSH, six 10 min. programs, 1993)
5. *Homeostasis* (FSH, 20 min., 1993)
6. *Homeostasis* (IM, 60 min., 1990)
7. *The Incredible Human Machine* (CBS, 60 min., 1992)
8. *Landscapes and Interiors* (FHS, 26 min., 1990)
9. *The Origin of Life* (EBF, 11 min., 1972)

10. *The Origin of Life* (FHS, 60 min., 1995)

11. *The Physics and Physiology of Sports* (FSH, 23 min., 1995)

12. *Your Body Video Series* (CA, VHS or filmstrip)

Videodiscs

1. *Anatomy and Physiology Videodisc* (VI, 1994)

2. *Atoms to Anatomy* (VI, 1994)

3. *Human Biology Videodisc* (FSH, 1995)

4. *Life Sciences Videodisc* (ODC, 1988)

5. *Slice of Life Videodisc* (EHSL, 1992)

6. *The Living Body* (FSH, 1992)

Computer Software

1. *A.D.A.M. Software* (ADAM)

2. *Interactive Atlas of Human Anatomy* (CIBA, CD ROM for WIN/MAC, 1996)

3. *Medworks* (VT, CD ROM for WIN/MAC, 1996)

2 Basic Chemistry

SUGGESTED LECTURE OUTLINE

I. CONCEPTS OF MATTER AND ENERGY (pp. 22–23)
   A. Matter (p. 22)
   B. Energy (pp. 22–23)
      1. Forms of Energy
      2. Energy from Conversions

II. COMPOSITION OF MATTER (pp. 23–28)
   A. Elements and Atoms (pp. 23–25)
   B. Atomic Structure (pp. 25–26)
      1. Planetary and Orbital Models of an Atom
   C. Identifying Elements (pp. 26–28)
      1. Atomic Number
      2. Atomic Mass Number
      3. Atomic Weight and Isotopes

III. MOLECULES AND COMPOUNDS (p. 29)

IV. CHEMICAL BONDS AND CHEMICAL REACTIONS (pp. 29–37)
   A. Bond Formation (pp. 29–36)
      1. Role of Electrons
      2. Types of Chemical Bonds
         a. Ionic Bonds
         b. Covalent Bonds
         c. Hydrogen Bonds
   B. Patterns of Chemical Reactions (pp. 36–37)
      1. Synthesis Reactions
      2. Decomposition Reactions
      3. Exchange Reactions

V. BIOCHEMISTRY: THE CHEMICAL COMPOSITION OF LIVING MATTER (pp. 37–50)
   A. Inorganic Compounds (pp. 37–40)
      1. Water
      2. Salts
      3. Acids and Bases
         a. Characteristics of Acids
         b. Characteristics of Bases
         c. pH: Acid-Base Concentrations
   B. Organic Compounds (pp. 40–50)
      1. Carbohydrates
         a. Monosaccharides
         b. Disaccharides
         c. Polysaccharides
      2. Lipids
         a. Neutral Fats
         b. Phospholipids
         c. Steroids

3. Proteins
    a. Fibrous and Globular Proteins
    b. Enzymes and Enzyme Activity
4. Nucleic Acids
5. Adenosine Triphosphate (ATP)

LECTURE HINTS

1. Emphasize the fact that energy is not created or destroyed, but can be converted from one form to another. Use food as an example, starting with the sun and ending with the chemical bonds of an ATP molecule.

2. Make sure that students understand the difference between the following terms: atom, element, compound, and molecule.

3. Use as many visual aids as you can to demonstrate bonding patterns between atoms. Ball-and-stick models or cutouts for the board will work well.

4. Make sure students understand that hydrogen bonds can be formed between molecules or can be formed within a molecule. Use water as an example of intermolecular hydrogen bonding and proteins for intramolecular bonding.

5. Construct a flowchart to help students identify molecular formulas for carbohydrates, lipids, and proteins. For example, if H:O = 2:1, the molecule is a carbohydrate. Then, if the carbohydrate has 3–7 carbons and exists in a single ring or a single chain, it is a monosaccharide. Then, put cutout pictures of various examples of carbohydrates, lipids, or amino acids on the board and ask students to classify them using the chart.

6. Many students learn that proteins are important because they are enzymes. Emphasize all the roles that proteins play in the human body.

7. Use a necklace of pop-it beads to demonstrate the importance of an enzyme's active site. Construct the necklace so that all the beads are of one color except for three or four of them. These beads will be the active sites. Use the necklace to show the effects of denaturation or mutation on protein function.

ANSWERS TO TEXTBOOK REVIEW QUESTIONS
(Questions appear on pp. 53–54)

1. Chemistry is basic to an understanding of anatomy and physiology because our bodies, everything in our environment, and everything we take into our bodies are all chemical substances.

2. Your body is composed of matter.

3. Energy has no mass and does not occupy space. It can only be defined by its effect on matter. Energy is defined as the ability to do work or put matter into motion.

4. Chewing food—mechanical energy; vision—light energy and electrical energy (of a nerve impulse); bending the fingers—mechanical energy; breaking ATP bonds—chemical energy.

5. Energy is not really lost in energy transformations, but some of it is unusable, being converted to heat.

6. Both lead and gold are elements. Elements are the unique building blocks of matter that cannot be transformed into one another.

7. Carbon (C), hydrogen (H), nitrogen (N), and oxygen (O) make up the bulk of living matter. Nitrogen is found chiefly in proteins and nucleic acids.

8. See Table 2.3, p. 28 of the text for answers.

9. An atom is the building block of an element or the smallest particle of an element that displays its properties.

10. All atoms contain equal numbers of protons and electrons.

11.
Particle	Position in Atom	Charge	Mass
Proton	Nucleus	+	1 amu
Neutron	Nucleus	None	1 amu
Electron	Orbital around the nucleus	-	0

12. Isotopes are different atomic forms of the same element that differ only in the number of neutrons they contain.

13. The heaviest isotopes of elements are often unstable and most likely to be radioisotopes. Spontaneous deterioration of an (unstable) atom by emission of particles or energy from the nucleus = radioactivity.

14. Electrons; valence shell.

15. A molecule is a combination of two or more atoms held together by chemical bonds. A molecule of an element contains chemically bonded atoms of the same element; a molecule of a compound contains chemically bonded atoms of different elements.

16. Ionic bonds form when electrons are transferred from one atom to another. In covalent bonds, electrons are shared among the atoms of the molecule.

17. Hydrogen bonds are fragile bonds formed when hydrogen atoms bonded to one molecule (or one part of a single molecule) are attracted by electron-hungry atoms of other molecules (or other atoms of the same molecule).

18. A polar molecule is formed only when the atoms of the molecule are different and have different electron-attracting abilities. Oxygen gas contains two oxygen atoms and is therefore nonpolar.

19. $2 Hg + O_2 \rightarrow 2 HgO$; synthetic.

    $Fe^{++} + CuSO_4 \rightarrow FeSO_4 + Cu^{++}$; exchange.

    $HCl + NaOH \rightarrow NaCl + H_2O$; exchange.

    $HNO_3 \rightarrow H^+ + NO_3^-$; decomposition

20. Inorganic compounds are noncarbon-containing: water, salts, acids, and bases. Organic compounds are carbon-containing: carbohydrates, lipids, proteins, and nucleic acids.

21. Water absorbs and releases large amounts of heat without changing appreciably in temperature; it is a polar molecule and an excellent solvent for chemical reactions and transport of substances in the blood and other body fluids; it is an important reactant in some chemical reactions; it is the basis of lubricants; water cushions and protects certain body organs (the brain).

22. An electrolyte is a chemical substance that conducts an electrical current when in solution.

23. Acids release hydrogen ions in water and are proton donors. Bases are proton acceptors.

24. pH is a measure of hydrogen ion concentration in solution. Distilled water is neutral; vinegar (acetic acid) is acidic; sodium bicarbonate is basic; gastric juice containing HCl is acidic.

25. Blood is slightly basic.

26. A pH of 3.3 is ten times more acidic than a pH of 4.3.

27. Monosaccharides = simple sugars, units of carbohydrates; examples are glucose, fructose, and galactose.

    Disaccharides = two simple sugars bonded together; examples are sucrose, lactose, and maltose.

    Polysaccharides = many simple sugars bonded together; examples are starch and glycogen.

    Carbohydrates are the major energy fuel for producing ATP in body cells.

28. Neutral fats consist of three fatty acid chains united to a glycerol backbone; they insulate and cushion the body and provide stored energy fuel. Phospholipids consist of two fatty acids and a phosphorus-containing group attached to glycerol; they are part of all cell membranes. Steroids are flat structures made of four interlocking rings. They, particularly cholesterol, form the basis of steroid hormones, vitamin D, and bile salts.

29. The R-group differs in the different amino acids, giving each type unique properties.

$$H_2N—\overset{\displaystyle R}{\underset{\displaystyle H}{C}}—COOH$$

30. Structural proteins are basically fibrous or linear proteins that have secondary structure; keratin, collagen, and elastin are examples. Functional proteins are globular or spherical proteins (having tertiary structure); examples are enzymes, some hormones, antibodies, and hemoglobin.

31. An enzyme is a biological catalyst. Enzymes increase the rate of chemical reactions by holding the reaction molecules in the proper position(s) to interact.

32. High body temperature and acidosis interfere with enzyme activity by destroying the three-dimensional structure that is essential for binding substrate and catalysis.

33. The structural unit of nucleic acids is the nucleotide. The major classes of nucleic acids are DNA and RNA. DNA is a double-stranded helix of nucleotides; its bases are A, G, C, and T, and its sugar is deoxyribose. RNA is single-stranded; its bases are A, G, C, and U, and its sugar is ribose. DNA is the genetic material that carries instructions for protein synthesis; RNA carries out DNA's instructions.

34. ATP is a modified nucleotide that is the immediate source of chemical energy for all body cells.

ANSWERS TO AT THE CLINIC QUESTIONS

1. If the antibiotic binds to the enzyme region that normally bonds to reactants, the bacterial enzyme's function will be blocked. Therefore, the bacteria may die and the person may get better.

2. a. pH is defined as the measurement of the hydrogen ion concentration in a solution. The normal blood pH is 7.4.

   b. Severe acidosis is critical because blood comes in contact with nearly every body cell and can adversely affect the cell membranes, the function of the kidneys, muscle contraction, and neural activity.

3. Neutral fats are found in deposits under the skin and surrounding organs. One of their functions is to insulate the body and prevent heat loss. Evelyn is thin and feels cold because she loses heat more readily than Barbara, who retains heat.

4. The doctor will probably use sonography (ultrasound imaging) because it exerts no known harmful effects on living tissue and is effective for viewing a fetus.

CLASSROOM DEMONSTRATIONS AND STUDENT ACTIVITIES

Classroom Demonstrations

1. Film(s) or other AV materials of choice.

2. Demonstrate the structure of simple molecules with ball-and-stick models.

3. Obtain two strings of dissimilar pop-it beads. Put the beads together to demonstrate a synthesis reaction, and take them apart to demonstrate a decomposition reaction. Take a bead from each different chain and put them together to illustrate an exchange reaction.

4. Use a Slinky to demonstrate denaturation of an enzyme. Tie colored yarn on the Slinky at two sites that are widely separated, and then coil and twist the Slinky upon itself to bring the two pieces of yarn next to each other. Identify the site where the yarn pieces are at the active site. Then remind students that when the hydrogen bonds holding the enzyme (or functional protein) in its specific 3-D structure are broken, the spherical structure (and the active site) is destroyed, by allowing the Slinky to resume its helical shape. (Best done when protein and enzyme structure has been discussed.)

5. Use a 3-D model of a portion of DNA to demonstrate its units and overall structure.

Student Activities

1. Ask students to name all the foods containing saturated fats and all those containing unsaturated fats that they have eaten in the last 24 hours.

2. Bring in materials or objects that are composed of common elements, e.g., a gold chain, a piece of coal, a piece of copper pipe, or a cast iron skillet. Also provide examples of common compounds, such as water, table salt, vinegar, sodium bicarbonate. Ask students to define atom, element, and compound, and explain how an atom and a molecule of a compound differ.

3. Provide wide-range pH paper and beakers containing (a) distilled water, (b) vinegar, (c) sodium bicarbonate solution, and (d) aqueous solutions of egg white, lemon juice, and ammonia. Ask students to determine the pH of each solution and to indicate whether it is acidic, neutral, or basic. Use this exercise as a lead-in to a discussion of common acids and bases and the importance of pH homeostasis in the body.

4. Student assignment for class discussion: Find examples of the uses of radioisotopes in popular articles or nursing journals, and be prepared to discuss the advantages and disadvantages of radioisotope use in medicine.

AUDIOVISUAL AIDS

Videotapes

1. *Atoms and Molecules* (AIMS Media, 30 min., 1991)

2. *The Atom Revisited* (FHS, 50 min., 1996)

3. *Basic Chemistry for Biology Students* (HRM, 30 min., 1993)

4. *The Chemistry of Life* (EI, VHS or slides)

5. *The Chemistry of Carbohydrates and Lipids* (EL, VHS or slides)

6. *The Chemistry of Nucleic Acids* (EL, VHS or slides)

7. *The Chemistry of Proteins* (EL, VHS or slides)

8. *DNA: The Master Molecule of Life* (EL, VHS or filmstrip)

9. *Enzymes: Regulators of Body Chemistry* (IM, 40 min., 1982)

10. *Francis Crick: Beyond the Double Helix* (CBS, 28 min., 1990)

11. *Functional Chemistry in Living Cells* (HRM, 1984)

12. *Radiation and You* (HV, 21 min.)

13. *Translating the Code: Protein Synthesis* (HRM, 30 min.)

14. *Unlocking the Secrets of Life* (FHS, 50 min., 1996)

15. *The World of Chemistry* (AC, 1989)

Videodisc

1. *Chemistry at Work* (VI, 1994)

Computer Software

1. *Biochemistry* (IM, CD ROM for WIN/MAC, 1995)

2. *DNA—The Master Molecule* (EME, Disk for MAC/IBM, 1993)

3   Cells and Tissues

SUGGESTED LECTURE OUTLINE

Part 1: Cells (pp. 56–76)

  I. OVERVIEW OF THE CELLULAR BASIS OF LIFE (p. 56)

 II. ANATOMY OF A GENERALIZED CELL (pp. 57–63)
   A. The Nucleus (pp. 57–58)
      1. Nuclear Membrane
      2. Nucleoli
      3. Chromatin
   B. The Plasma Membrane (pp. 58–59)
      1. Specializations of the Plasma Membrane
   C. The Cytoplasm (pp. 59–63)
      1. Cytoplasmic Organelles
         a. Ribosomes
         b. Endoplasmic Reticulum
         c. Golgi Apparatus
         d. Lysosomes
         e. Peroxisomes
         f. Mitochondria
         g. Cytoskeleton
         h. Centrioles

 III. CELL PHYSIOLOGY (pp. 63–76)
   A. Membrane Transport (pp. 63–69)
      1. Passive Transport Processes: Diffusion and Filtration
         a. Diffusion
         b. Filtration
      2. Active Transport Processes
         a. Solute Pumping
         b. Bulk Transport
   B. Cell Division (pp. 69–73)
      1. Preparations: DNA Replication
      2. Events of Cell Division
         a. Mitosis
         b. Cytokinesis
   C. Protein Synthesis (pp. 73–76)
      1. Genes: The Blueprint for Protein Structure
      2. The Role of RNA
      3. Transcription
      4. Translation

Part 2: Body Tissues (pp. 76–88)

  I. EPITHELIAL TISSUE (pp. 76–80)
   A. Special Characteristics of Epithelium (p. 77)

B. Classification of Epithelium (pp. 77–80)
   1. Simple Epithelia
     a. Simple Squamous Epithelium
     b. Simple Cuboidal Epithelium
     c. Simple Columnar Epithelium
     d. Pseudostratified Columnar Epithelium
   2. Stratified Epithelia
     a. Stratified Squamous Epithelium
     b. Stratified Cuboidal and Stratified Columnar Epithelia
     c. Transitional Epithelium
   3. Glandular Epithelium

II. CONNECTIVE TISSUE (pp. 80–83)
  A. Common Characteristics of Connective Tissue (p. 80)
  B. Types of Connective Tissue (pp. 80–83)
   1. Bone
   2. Cartilage
   3. Dense Connective Tissue
   4. Areolar Tissue
   5. Adipose Tissue
   6. Reticular Connective Tissue
   7. Blood

III. MUSCLE TISSUE (pp. 83–85)
  A. Types of Muscle Tissue (pp. 83–85)
   1. Skeletal Muscle
   2. Cardiac Muscle
   3. Smooth Muscle

IV. NERVOUS TISSUE (p. 85)

V. TISSUE REPAIR (WOUND HEALING) (pp. 86–87)

VI. DEVELOPMENTAL ASPECTS OF CELLS AND TISSUES (pp. 87–88)

LECTURE HINTS

1. Emphasize that the cell described in this chapter is an "idealized cell" and that there are many cells in the body that have a different structure. Mature red blood cells, for example, are anucleate, while skeletal muscle cells are multinucleate and are elongated. Then emphasize that it is the specialized function of these cells that makes this structure preferable.

2. Show the students slides of electron micrographs of cells to augment the diagrams that appear in the text. This is a good opportunity to discuss how animal tissues are prepared for microscopy and to discuss different types of microscopes.

3. The terms hypertonic, hypotonic, and isotonic should be broken into their roots so that their meaning is immediately obvious to the student. For example, hypertonic means great pulling power or great tension; therefore, water is drawn into a hypertonic solution. Hypotonic, on the other hand, means low pulling power, or low tension. Therefore, hypotonic solutions lose water to hypertonic solutions. Students might want to envision a tug-of-war game, with a water molecule in the middle. Provide lots of problems for students so that these terms become part of their vocabulary.

4. Although the stages of the cell cycle are described as discrete events, students should be shown that they represent a continuous process. Films or videodisc segments showing the cell cycle will help students visualize this better. Students should realize that interphase is important and is a period of intense activity. Ask students to discuss why telophase is the reverse of prophase.

5. Make sure that students understand the relationship between genes, chromosomes, DNA, and proteins.

6. Students will have trouble identifying tissues when given microscope slides or micrographs. Encourage them to identify the tissue based on tissue characteristics they have learned. Students could construct a flowchart for this purpose. For example, if cells are very close with little or no extracellular material, it is epithelium not connective tissue. If it is multilayered, it is stratified epithelium. If it is a single layer, it is simple epithelium, etc.

7. This is an excellent time to pique student interest by discussing topics such as the cell biology of cancer or theories of aging.

ANSWERS TO TEXTBOOK REVIEW QUESTIONS
(Questions appear on pp. 91–92)

1. Oxygen, carbon, nitrogen, and hydrogen.

2. Organelle: Literally, little organ. An intracellular structure that performs a specific function for the cell.

   Cell: The basic living unit of structure and function.

3. All cells are able to metabolize, divide, grow, respond to stimuli, digest nutrients, and excrete wastes.

4. The nucleus is the control center of the cell. DNA is the storehouse of the cell's genetic instructions, which specify protein structure. DNA is contained in the chromatin. Nucleoli are sites of ribosome formation.

5. The plasma membrane is composed of a double layer of lipids (phospholipids and cholesterol), with proteins floating in the lipid. Some of the proteins have attached sugar groups. The plasma membrane serves as a semipermeable barrier that contains cell contents, and functions in membrane transport and in cell-to-cell interactions.

6. The cytoplasm is the "factory region" of the cell; that is, the general, daily activities of the cell are carried out here (protein synthesis, ATP synthesis, and the like). Cytosol, organelles, and inclusions comprise the cytoplasm.

7. Ribosomes: Protein synthesis.

   Endoplasmic reticulum: Intracellular transport of proteins made on the ribosomes (rough ER) or formation of lipids/steroids.

   Golgi apparatus: Packaging of proteins for export from the cell.

   Lysosomes: Breakdown of "worn-out" cell organelles or ingested foreign materials, such as bacteria.

   Mitochondria: The major sites of ATP synthesis in the cell.

   Centrioles: Structures that "spin" the mitotic spindle during cell division; form bases of cilia and flagella.

   Peroxisomes: Detoxify harmful chemicals, such as alcohol and oxygen-free radicals.

Cytoskeleton: Formed of microtubules and different types of filaments that construct the internal framework of the cell and promote cellular movements.

8. An active transport process requires that the cell use ATP to energize the process. A passive transport process is energized by kinetic energy or pressure differences.

9. Diffusion: The movement of particles from an area of higher concentration to an area of lower concentration as a result of their kinetic energy; a passive process.

Osmosis: The diffusion of water through a semipermeable or selectively permeable membrane.

Simple Diffusion: The unassisted diffusion of solutes through a semipermeable or selectively permeable membrane.

Filtration: The passage of solutes and solvent through a membrane from an area of higher hydrostatic pressure to an area of lower hydrostatic pressure.

Solute pumping: The movement of substances across a membrane by a solute pump, a "carrier" protein present in cell membrane. Requires that ATP be used and usually occurs against concentration and electrical gradients.

Exocytosis: A mechanism by which substances are moved from the cell interior to the extracellular space as a vesicle fuses with the plasma membrane.

Endocytosis: A means by which fairly large extracellular molecules or particles enter cells.

Phagocytosis: The engulfment of particulate (solid) substances by a cell through the cell's forming pseudopods (false feet).

Pinocytosis: The internalization of soluble substances (proteins and the like) by a cell through vesicle formation.

10. Passive process: The size of the pores and whether the substance is soluble in the lipid (fat) portion of the membrane.

Active process: Whether the proper carrier proteins (pumps) are present in the membrane and in what amounts.

11. Hypertonic solutions cause cells to become crenated (shrunken) as water leaves by osmosis. Hypotonic solutions cause cells to become bloated as water enters them from an area of higher water concentration. If there is a substantial difference in water concentration, the cell will burst as excessive water enters it. Isotonic solutions have the same water/solute concentrations as cells; thus no structural changes occur.

12. The DNA helix uncoils and the hydrogen bonds holding the bases together are broken by enzymes. Each freed nucleotide strand then acts as a model for building a complementary strand from DNA nucleotides.

13. Mitosis: Nuclear division. Mitosis is important because it provides cells needed for growth and body repair.

14. The spindle acts as a scaffolding or attachment site for the chromosomes during mitosis.

15. If the cells of an organ are amitotic, the cells that repair any damage to that organ will restore its structure but not necessarily its ability to function. For example, when heart cells die, they are replaced by connective tissue cells (scar tissue), which are unable to contract as do heart muscle cells. Likewise, damaged nerve cells are replaced by scar tissue, which is unable to transmit electrical signals. When tissue repair occurs by regeneration (mitosis of the same type of cells), normal function is restored.

16. DNA, the "genes," provides the specific instructions for building proteins; that is, each three-base sequence along the gene indicates the precise amino acid that is to appear in the protein at that relative position. RNA carries out DNA's instructions and sees that the protein is built. Messenger RNA carries the "message" to the ribosome. Transfer RNA transports amino acids to the ribosome, and ribosomal RNA forms part of the ribosome, or protein synthesis site.

17. Tissue: A group of cells similar in structure and function. The four major tissue types are epithelial, connective, muscle, and nervous. Connective tissue is the type most widely distributed in the body.

18. The most general characteristic of epithelial tissue is that it lines body and organ cavities, covers the body surface, and forms glands. It is avascular and regenerates well. Its most important functions are protection, absorption, filtration, and secretion. For example, the external epithelium (epidermis) protects against bacterial, chemical, and thermal damage; that of the respiratory system has cilia, which sweep debris away from the lungs; and that of the digestive system is able to absorb substances.

19. On the basis of cell shape (squamous, cuboidal, or columnar) and the relative number of cell layers; that is, one cell layer is a simple epithelium, and more than one layer is a stratified epithelium. Stratified epithelia are named according to the cell type found in the top layer.

20. Ciliated epithelium is found in the respiratory system, where it acts to prevent debris from entering the lower respiratory passageways, and in the reproductive system, where it acts to move sex cells along the duct passageways.

21. Endocrine glands are ductless; their product is always a hormone. Exocrine glands have ducts, which conduct the glandular product to an epithelial surface; the products are varied, for example, sweat, pancreatic hormones, liver bile.

22. Connective tissues are usually characterized by a large amount of extracellular *matrix*, which is secreted by the living cells. Connective tissues serve to connect, support, protect, and repair other body tissues. The functions of connective tissue are best explained by its matrix; that is, when connective tissue is a cushioning tissue, the matrix is soft and pliable; when the connective tissue is meant to support or give strength to the body, the matrix is hard/strong.

23. a. Areolar; b. Bone.

24. To contract or shorten.

25. Skeletal: Attached to the skeleton and providing for gross body movement.

    Smooth: Found in the walls of internal organs and providing for the movement of substances (for example, urine, food) through internal body tracts.

    Cardiac: Forms the heart, which propels blood through the blood vessels.

26. When a muscle type is said to be involuntary in action, this means that one cannot voluntarily control its action. Skeletal muscle is said to be voluntary in action.

27. Irritability (ability to respond to a stimulus) and conductivity (ability to transmit an impulse).

28. Neurons, like other cells, have a nucleus, cytoplasm, cell membrane, and the usual organelles. Neurons differ from other cells in that they are usually extremely elongated. This difference reflects their special function, which is to conduct impulses over long distances in the body.

29. Neoplasm: Growth or cell mass that results from the loss of the normal controls of cell division.

    Atrophy: A decrease in the size of a body organ or body part as a result of the loss of normal stimulation (exercise or nerve stimulation).

    Hyperplasia: An increase in the size of a body organ that results from a local stimulant; usually temporary.

30. Benign neoplasms tend to remain localized and encapsulated; they may cause harm to the body by exerting pressure on neighboring organs. Cancerous neoplasms tend to be uncontrolled and spread to distant body areas, where they proliferate and cause damage.

ANSWERS TO AT THE CLINIC QUESTIONS

1. The presence of a greater-than-normal concentration of solutes that are not absorbed (bacterial toxins, cell breakdown products, or inflammatory chemicals) will "hold" water in the intestine because of its high osmotic pressure. As a result, more water leaves the body in feces, causing diarrhea.

2. Vincristine: Anything that damages the mitotic spindle interferes with cell division and, hence, would prevent proliferation of the cancer cells.

    Adriamycin: If messenger RNA cannot be made, proteins cannot be made either. All cells must have essential enzymes to function properly.

3. Lysosomal destruction releases acid hydrolases into the cytoplasm, killing the cell. When the cell lyses, inflammation is triggered. Glucocorticoids—of which hydrocortisone is one—control the unpleasant effects of inflammation by decreasing edema.

4. Cartilage heals slowly because it lacks the blood supply necessary for the healing process.

5. Only the liver will fully recover because it is composed of epithelial tissue, which completely regenerates. The injured areas of the heart and the brain grow back as scar tissue and thus do not completely regenerate.

6. Hyperplasia refers to a condition in which cells have been stimulated to divide, resulting in an increase in cell number and an enlarged organ or tissue. Dysplasia is a condition in which the cells of a tissue abnormally vary in size, shape, and organization. Neoplasia means a new growth. It results from increased cell division that is autonomous and outside of the influence of normal regulation and is characteristic of cancer cells. Kareem does not have cancer of the mouth.

CLASSROOM DEMONSTRATIONS AND STUDENT ACTIVITIES

Classroom Demonstrations

1. Film(s) or other AV materials of choice.

2. Use an animal cell model to demonstrate the various organelles and cell parts.

3. Use models of epithelial and connective tissue and muscle cell types, and of a neuron, to illustrate how the cells of the different tissue types are both similar and dissimilar.

4. Use models of the events of mitosis to support your class presentation of cell division

5. Secure a glass funnel containing filter paper over a beaker. Illustrate how greater fluid pressure (provided by more fluid in the funnel) leads to faster filtration.

6. Set up three microscopes with blood cells suspended in (a) physiologic saline, (b) hypertonic saline, and (c) distilled water for class observation to illustrate the influence of solutions of different tonicity on living cells as you explain the terms *isotonic*, *hypertonic*, and *hypotonic*.

7. Set up a simple diffusion demonstration as follows: Obtain a Petri plate with solidified agar-agar, and place a large crystal of some histological dye (potassium permanganate, methylene blue, etc.) on its surface. Cover and let stand for a few hours before class. The diffusion ring with the deepest color closest to the crystal will provide clear evidence of the phenomenon of diffusion. Alternatively, if an overhead projector is available, put the projector on and place a dye crystal in a beaker of water on the projector. An image of the diffusing stain can then be projected on the screen. (This demonstration is quite exciting to most students.)

8. Set up a simple osmometer before class and have students observe the fluid level in the tube from time to time. (Put glucose solution in a dialysis sac, and tie the sac tightly to a length of glass tubing. Secure the glass tubing to a ring stand with a clamp; the glucose-containing dialysis sac should be immersed in distilled water in a beaker.) Use this demonstration to support your discussion of osmosis.

9. You can make models of epithelial cells by using bricks and tiles of various sizes and adding felt nuclei, cilia, etc.

Student Activities

1. Ask students to explain how the plasma membrane is structurally like a chocolate bar with almonds, or how the cell itself is like fruit gelatin.

2. To explore the fat-protein makeup of the plasma membrane, tell students that chloroform, ether, and most anesthetics are fat-soluble substances. Ask them to explain why this chemical characteristic is important.

3. Have the students name all the common examples of diffusion, osmosis, and filtration in their daily lives that they can think of.

4. Provide colored pieces of yarn or pipe cleaners for students to manipulate to demonstrate the four phases of mitosis.

5. Ask students to make a list of all of the things their bodies could not do if they had no connective tissue.

6. Make cards for the blackboard listing the events of the cell cycle (chromosomes appear, spindle breaks down, etc.) and the stages of the cell cycle (interphase, prophase, etc.). Attach the cards to the board in a scrambled fashion using magnets or tape. Have students group the events under the proper heading, then have them place them in the order in which they occur under each heading.

7. Students should ponder the following questions and be prepared to answer them at the next class meeting:

   a. Why is precise division of the chromosomes during metaphase of mitosis so important?

   b. Why is damage to the heart or the brain more damaging than injury to the liver, for example?

   c. Is mitosis without cytokinesis possible? What would the result be?

8. Show diagrams or overheads of specialized human cells, such as sperm cells, or cells with microvilli. Ask students to make guesses about the function of those cells based on their structure.

AUDIOVISUAL AIDS

Videotapes

1. *Aging* (FHS, VHS or videodisc, 28 min., 1991)
2. *Architecture of Cells: Special Structure, Special Function* (HRM, 45 min., 1987)
3. *The Biochemical Basis of Biology—Cell Structure and Energy Production* (CBS, 60 min., 1992)
4. *The Biochemical Basis of Biology—DNA and Protein Synthesis* (CBS, 60 min., 1992)
5. *The Cell* (IM, 29 min., 1990)
6. *Cell Biology* (CIF, 15 min., 1989)
7. *Cell Division and the Life Cycle* (HRM, 1991)
8. *Cell Structure and Function* (EIM)
9. *Century 3: The Gift of Life* (USMAC, 28 min., 1975)
10. *Cytology and Histology* (EIL, VHS or slides)
11. *The Genetic Code: Transcription and Protein Synthesis* (IM, 13 min., 1992)
12. *An Introduction to Cells* (EIL, 60 min.)
13. *An Introduction to Living Cells* (IM, 20 min., 1992)
14. *Introduction to the Living Cell* (IM, 35 min., 1996)
15. *The New Cell* (IM, 40 min., 1990)
16. *Origin of Cellular Life* (IM, 40 min.)
17. *The Plasma Membrane* (IM, 15 min., 1988)
18. *Translating the Code: Protein Synthesis* (HRM, 25 min., 1992)

Videodisc

1. *Cell Biology* (CBS, 1992)

Computer Software

1. *Cell Biology* (IM, CD ROM for MAC/WIN/DOS, 1995)
2. *Enzyme Investigations* (EME, CD ROM for WIN/MAC or disk for WIN/MAC/Apple, 1992)
3. *Osmosis Lab* (EME, CD ROM for WIN/MAC or disk for WIN/MAC/Apple, 1994)

4   Skin and Body Membranes

SUGGESTED LECTURE OUTLINE

  I. CLASSIFICATION OF BODY MEMBRANES (pp. 94–96)
      A. Epithelial Membranes (pp. 94–96)
          1. Cutaneous Membrane
          2. Mucous Membranes
          3. Serous Membranes
      B. Connective Tissue Membranes (p. 96)

 II. INTEGUMENTARY SYSTEM (SKIN) (pp. 96–109)
      A. Basic Skin Functions (pp. 96–97)
      B. Structure of the Skin (pp. 97–100)
          1. Epidermis
          2. Dermis
      C. Skin Color (pp. 100–101)
      D. Appendages of the Skin (pp. 101–104)
          1. Cutaneous Glands
              a. Sebaceous (Oil) Glands
              b. Sweat Glands
          2. Hairs and Hair Follicles
          3. Nails
      E. Homeostatic Imbalances of the Skin (pp. 104–109)
          1. Infections and Allergies
          2. Burns
          3. Skin Cancer
              a. Basal Cell Carcinoma
              b. Squamous Cell Carcinoma
              c. Malignant Melanoma

III. DEVELOPMENTAL ASPECTS OF SKIN AND BODY MEMBRANES (p. 109)

LECTURE HINTS

1. Show micrographs or microscope slides of various types of skin sections. Emphasize that the diagrams of skin in the text are idealized pictures, and that skin varies at different regions of the body with age, sex, and race.

2. Students learn in this chapter that burns are classified according to depth and/or surface area covered. Emphasize that there are other factors to consider, such as type of burn (electrical, chemical, etc), area of the body burned (chest, face, hands, etc.), and age. When discussing the consequences of serious burning, refer to the functions of the integument. Loss of skin leads to loss of protection, loss of water, etc.

3. Use recent popular articles to help students appreciate the effects that nutrition, stress, fatigue, smoking, and sunlight have on the skin.

4. Students will be interested in discussing the skin as a thermoregulatory organ. Ask students to discuss changes in their skin on hot and cold days, or when they have a fever.

Remind them of behavioral changes as well, such as curling up under a blanket when sleeping on a cold night or stretching out without a cover on a hot night.

5. When discussing the production and growth of hair, students will be interested in hair treatment, such as permanents and straightening procedures. This is an opportunity to review protein structure and intramolecular bonding. Remember when beauty salons smelled like sulfur? Hair removal processes, such as electrolysis, and hair color could be discussed.

6. Use a variety of diagrams to help students understand the terms visceral and parietal.

7. As a point of interest, note that although vitamin D is produced in the skin, it is activated by the kidney. Thus, premature infants with unstable kidneys are at risk for developing rickets.

ANSWERS TO TEXTBOOK REVIEW QUESTIONS
(Questions appear on p. 112)

1. A mucosa is an epithelial lining of body cavities that are open to the exterior, e.g., the lining of the respiratory, digestive, and urinary tracts. A serosa lines body cavities that are closed to the body exterior, for example, pleura, pericardium, and peritoneum.

2. Synovial membrane.

3. Epithelium (epidermis) and connective tissue (dermis).

4. Chemical (e.g., acids and bases), mechanical (pressure, trauma), bacterial (via the acid mantle), desiccation (via the waterproofing keratin), and thermal.

5. When the amount of UV radiation reaching the skin increases (as when sitting in the sun), the melanocytes in the skin begin to produce increased amounts of protective melanin (a brown pigment) and to deposit it in epidermal skin layers to protect the basal cells from the damaging effects of UV radiation.

6. Decubitus ulcer: A localized breakdown of the skin and underlying tissues resulting from a lack of blood supply to the area.

7. Perspiration (water, salt, and sometimes protein/lipid substances) from sudoriferous glands; sebum (fatty substances and broken cells) from sebaceous glands.

8. All temperature-regulating functions of the skin are controlled by the nervous system. Under conditions of high external temperature, blood is allowed to flush into the skin capillary net so that heat radiates from the body surface. At the same time, sudoriferous glands produce perspiration, which evaporates from the skin surface so that more body heat is dissipated. With low external temperatures, blood is prevented from entering the skin capillary system, and thus radiation to the body exterior is prevented and perspiration does not occur. This restricts the blood to deeper tissues and maintains the core temperature of the body.

9. Blackhead: An accumulation of dried sebum and bacteria in a sebaceous gland.

10. Arrector pili are small bands of smooth muscle cells attached to a hair follicle that pull the hair (follicle) into an upright position when they contract. They are activated by cold or fright.

11. The immediate problem is loss of body fluids containing essential proteins and electrolytes. Then, overwhelming bacterial infection places the body at risk.

12. First degree: Epithelial damage only; redness and swelling with some pain. Regenerates well.

Second-degree: Epithelial destruction and some dermal damage; same symptoms as first degree plus blisters form. Usually regenerates well.

Third-degree: Epidermal and dermal destruction; skin is blanched or blackened. No pain because nerve endings are destroyed. Must be grafted.

13. This is a result of a delayed-action gene that (eventually) causes the nonproduction of melanin.

14. It dries, thins (leading to the increased possibility of bruising), and loses elasticity and subcutaneous fat tissue.

15. No. Very fine hairs are formed but they are colorless and many do not actually emerge from the follicle.

ANSWERS TO AT THE CLINIC QUESTIONS

1. Cyanosis: A blue cast to the skin and mucosae. Its presence indicates a lack of oxygen in the blood.

2. Due to the lack of adipose tissue, there is a decrease in the skin's ability to act as a shock-absorber and insulator.

3. His long-term exposure to UV radiation in sunlight is considered to be a risk factor related to the development of skin cancer. In addition, moles or pigmented spots that show (A) asymmetry, (B) border irregularity, (C) color cariation, and (D) a diameter greater than 6 mm are all signs of a possible malignant melanoma. He should seek immediate medical attention. If it is a malignant melanoma, the chance for survival is not high, but early detection increases survival rates.

4. The Orange color might be due to carotene deposited in the stratum corneum and subcutaneous tissue. Carotene is found in orange vegetables such as carrots and in deep-yellow and leafy green vegetables.

5. Keratin found in cells of the stratum corneum is a water-repellant protein. Its presence prevents water intake from the environment.

CLASSROOM DEMONSTRATIONS AND STUDENT ACTIVITIES

Classroom Demonstrations

1. Film(s) or other AV materials of choice.

2. Demonstrate the structure of the skin using a 3-D model.

3. Thrust your fist into a flaccid balloon to demonstrate the relationship between the parietal and visceral layers of a serosa.

4. Use a human torso model to show the body location of the mucous and serous membranes.

Student Activities

1. Have a small fan operating. As students file into the classroom, spray their arm or hand with water (a simple window-cleaner spray bottle may be used as the "sprayer"). Ask them to describe their sensations as the water evaporates and to explain why evaporation of water (sweat) from the skin surface is important to body homeostasis.

2. Indicate to the class that many animals (snakes, insects, and lobsters, for instance) shed their skin periodically. Ask them to comment on how humans compare in this aspect.

3. Student assignment for class discussion:

   a. Look up the current drug treatment for baldness in the medical library (minoxidil).

   b. Find references in current journals to the damaging effects of sunlight on the skin, and be prepared to describe the difference in the A and B types of UV rays relative to skin damage.

4. Show students a picture of a heavily wrinkled person. Ask them to list all of the factors that might have contributed to the skin deterioration. Discuss the research that led to the use of retinoic acid for the treatment of photo-aged skin.

5. Provide magnifying lenses. Have the students use them to scrutinize their skin and the mucosa of a classmate, and then describe important differences visible in these two types of covering epithelia.

6. Provide small glass plates and instruct students to observe the change in the color of their skin while pressing the heel of their hand firmly against the glass. Ask them to explain the reason for the color change and what would happen to their skin cells if the pressure was prolonged.

7. Student assignment for class discussion: Look up the signs and symptoms of basal cell carcinoma and malignant melanoma for class discussion.

8. Student assignment for class discussion: Have students research and report on FDA regulations for the classification of cosmetics and drugs.

AUDIOVISUAL AIDS

Videotapes

1. *Burns and Heat Stroke* (PLP, 10 min., 1990)

2. *First Aid: Newest Techniques* (SC, 26 min., 1988)

3. *Skin* (FHS, 20 min., 1995)

4. *Skin Cancer* (PLP, 10 min., 1989)

5. *Skin Deep* (FHS, VHS or videodisc, 26 min., 1990)

6. *Skin and Hair* (EBF, 10 min., 1987)

7. *The Skin—Its Structure and Function* (EBF, 20 min., 1983)

5 The Skeletal System

C. Bones of the Pelvic Girdle (pp. 136–139)
D. Bones of the Lower Limbs (pp. 139–142)
   1. Thigh
   2. Leg
   3. Foot

IV. JOINTS (pp. 142–148)
   A. Fibrous Joints (pp. 142–143)
   B. Cartilaginous Joints (pp. 142–143)
   C. Synovial Joints (pp. 143–145)
   D. Inflammatory Disorders of Joints (pp. 144–148)

V. DEVELOPMENTAL ASPECTS OF THE SKELETON (p. 148)

LECTURE HINTS

1. Discuss with students the differences between the skeletons of males and females, athletes and sedentary people, and the changes in the skeleton that take place with normal aging. Heterotrophic bones and accessory bones may also be found in many people. A discussion of what information skeletons provide in forensic medicine would be interesting.

2. Mention that osteogenic sarcoma and leukemia are common cancers in people exposed to radiation because of the deposition of minerals in bone.

3. Emphasize that bone is not a static tissue, that it is vascular, and that it is constantly produced and reabsorbed for the purpose of calcium homeostasis and to allow it to accommodate functional and gravitational stress. Moderate weight-bearing exercise will stimulate bone formation and may help prevent diseases such as osteoporosis.

4. Students are surprised to hear that ossification is incomplete at birth. Consult a developmental biology book, and share a timetable for ossification with students.

5. Alert students to the risk factors for osteoporosis, and discuss current treatment and future therapies.

6. Carefully stand on a china cup to demonstrate the compressional strength that calcium/bone salts provide to bone tissue, and use a leather belt to demonstrate the tensile strength that collagen provides to bone tissue.

7. Use a flexible, articulated skeleton to demonstrate the movements permitted at diarthrotic joints.

ANSWERS TO TEXTBOOK REVIEW QUESTIONS
(Questions appear on pp. 151–152)

1. Forms the body's internal structural framework, that is, provides support. Anchors skeletal muscles and allows them to exert force to produce movement. Protects by enclosing (skull, thorax, and pelvis). Provides a storage depot for calcium and fats. Site of blood formation.

2. Long: All limb bones except those of the wrist and ankle, that is, the femur, fibula, tibia, humerus, ulna, radius, and phalanges.

   Short: Basically cube-shaped bones of the wrist and ankle.

   Flat: Bones of the skull, sternum. Irregular--os coxa, vertebrae.

3. Shaft of a long bone: Diaphysis.

    Ends of a long bone: Epiphyses.

    Yellow marrow: Fat.

    Spongy bone looks holey or lacey, whereas compact bone appears to be solid.

4. Bone is highly vascularized and thus heals rapidly. Cartilage has poor vascularization and depends on diffusion for its nutrient supply; thus, it heals slowly or poorly.

5. Organic part: Flexibility.

    Inorganic part (bone salts): Strength, hardness.

6. PTH promotes calcium homeostasis of the blood and is the most important factor determining if calcium is to be removed from or deposited in the bony skeleton. For example, when blood calcium levels begin to drop, PTH activates the osteoclasts of bone. As bone matrix is broken down, ionic calcium is released to the blood. Mechanical forces acting on bones determine where calcium can safely be removed or where more calcium salts should be deposited to maintain bone strength.

7. Fracture: A break in a bone. Compression and comminuted fractures are particularly common in the elderly. Greenstick fractures (incomplete fractures) are common when the bone matrix contains relatively more collagen, as do children's bones.

8. Skull, thorax, vertebral column.

9. Two each: Temporal and frontal bones.

    One each: Occipital, frontal, sphenoid, and ethmoid bones.

10. Frontal suture: Frontal and parietal bones.

    Sagittal suture: Parietal bones.

11. Temporomandibular joint.

12. Maxillary, frontal, sphenoid, and ethmoid bones.

13. Chin: Mandible.

    Cheekbone: Zygomatic.

    Upper jaw: Maxilla.

    Eyebrow ridges: Frontal.

14. The fetal skull has (a) much larger cranium-to-skull size ratio, (b) foreshortened facial bones, and (c) fontanelle or unfused (membraneous) areas.

15. Cervical, thoracic, lumbar, sacral, coccygeal.

16. See Figure 5.15 (p. 130) and Figure 5.17 (p. 131).

17. To cushion shocks to the head and to allow movement of the spinal column (e.g., laterally).

18. Sternum, ribs (attached to the vertebral column posteriorly).

19. True rib: Attached directly to the sternum by its own costal cartilage.

    False rib: Attached to the sternum indirectly (or not at all). A floating rib is a false rib. Floating ribs are easily broken because they have no anterior (sternal) attachment (direct or indirect) and thus have no anterior reinforcement.

20. Inverted cone.

21. Clavicle and scapula.

22. Humerus, radius, carpals.

23. Bearing weight (strength).

24. Ilium, ischium, pubis. The ilium is the largest. The ischium has the "sit-down" tuberosities. The pubis is most anterior.

25. The female pelvis is lighter, broader, and shallower and has a broader pubic arch and larger inlet and outlet.

26. Femur, fibula/tibia, tarsals, metatarsals, phalanges.

27. To connect bones and to allow movement (to a greater or lesser degree).

28. Synarthrotic: Essentially immovable.

    Amphiarthrotic: Slightly movable.

    Diarthrotic: Freely movable.

29. The articulating ends of bones in a synovial joint are separated by a cavity that contains synovial fluid. Synovial joints are enclosed by a fibrous connective tissue capsule lined with a smooth synovial membrane. Reinforcing ligaments may reinforce the fibrous capsule, and bursae may cushion tendons where they contact bone.

30. Arthritis: Inflammation of the joints. Degenerative arthritis (osteoarthritis) is the type most common in the elderly (owing to the wear and tear on joints that accompanies increasing age and use). Rheumatoid arthritis is believed to be an autoimmune disease.

31. Factors that keep bones healthy: physical stress/use (most important), proper diet (e.g., calcium).

    Factors that cause bones to become soft or atrophy: Disuse, hormone imbalances, loss of stimulation.

ANSWERS TO AT THE CLINIC QUESTIONS

1. The youngster had more organic material in her bones, allowing them to bend, while her grandmother's bones are completely calcified, having little organic material, and also probably thin due to osteoporosis.

2. No; the palatine bones are posterior to the palatine processes of the maxillae. If the palatine processes do not fuse, then the palatine bones remain unfused as well.

3. Most likely the paranasal sinuses on the right side of the face.

4. Most likely osteoporosis, a condition common in older women. A decline in bone mass, particularly in the spine and neck of the femur, increases the probability of fractures.

5. Dislocation. The head of the humerus has been forced out of its normal position in the glenoid cavity.

CLASSROOM DEMONSTRATIONS AND STUDENT ACTIVITIES

Classroom Demonstrations

1. Film(s) or other AV materials of choice.

2. Use an articulated skeleton to (a) indicate its protective and support aspects, (b) point out/identify individual bones, and (c) identify the security aspects of various body joints (e.g., sutures, hip vs. shoulder joint, hinge joints like the elbow and knee, etc.).

3. Use a prepared long bone (or purchase a beef bone from the market and have the butcher saw it longitudinally) to demonstrate the structural aspects of long bones. (The fresh bone is excellent for demonstrating the large difference between red and yellow marrow.)

4. Use a Beauchene (disarticulated) skull to demonstrate more clearly the individual skull bones and to show the fragile internal structure of bones containing sinuses (ethmoid, sphenoid, etc.).

5. Use a fetal skeleton to emphasize the changes in skull and body proportions that occur after birth, and to point out the fact that initially the skeleton is formed (mostly) of hyaline cartilage rather than bone.

6. Obtain X-rays of abnormal spinal curvatures (lordosis, scoliosis, etc.) to show the class.

7. Use a 3-D model of bone to identify the Haversian canals, lacunae containing osteocytes, lamellae, and canaliculi.

8. Obtain a video on the subject, or request that a local orthopedic surgeon visit the class to describe the techniques and advantages of arthroscopic joint surgery.

9. Use chicken bones that have been soaked in dilute acetic acid or baked in an oven to demonstrate how collagen and calcium/bone salts contribute to the flexibility and strength of bone.

Student Activities

1. If possible, remove an intervertebral disk (or its representative) from the articulated spinal column, and invite class discussion on what symptoms might be experienced if a disk ruptured (or pressed on nerve roots).

2. Using an articulated skeleton, quickly point out different individual bones. As a group, have students call out the names of the bones indicated.

3. Student assignment for class discussion:

   a. Use a nutrition reference to look up the roles of calcium in the body (other than making bones hard).

   b. Be prepared to discuss in some depth the various hormonal influences on the skeleton, and find out what drugs are currently used to help correct gigantism and dwarfism.

   c. Answer the question, "Why are incomplete or greenstick fractures more common in children than in adults?"

4. Have students research and report on current methods for the prevention and treatment of osteoporosis.

5. Using a disarticulated skeleton, hold up various bones and have students try to classify them as long, short, flat, or irregular.

6. Place a bone in a bag and have the students identify it using only their sense of touch.

AUDIOVISUAL AIDS

Videotapes

1. *Bones and Movement* (KSU, 9 min., 1990)

2. *Bones and Joints* (FHS, 20 min., 1995)

3. *The Development of Bone* (AGC, 30 min., 1975)

4. *Growth and Change* (FHS, 28 min., 1990)

5. *Locomotion and Skeletons* (IM, 29 min., 1978)

6. *Moving Parts* (FHS, 26 min., 1990)

7. *Osteoporosis: Progress and Prevention* (FHS, 24 min., 1995)

8. *The Skeleton* (EBF, 17 min., 1980)

9. *The Skeleton—An Introduction* (TFI, 46 min., 1990)

10. *Total Hip Arthroplasty—A New Approach* (EIL, 15 min., 1991)

Videodisc

1. *Nervous System, Muscular and Skeletal System* (NGS)

Computer Software

1. *The Interactive Skeleton* (IM, CD ROM for WIN/MAC, 1995)

6   The Muscular System

SUGGESTED LECTURE OUTLINE

I. OVERVIEW OF MUSCLE TISSUES (pp. 154–158)
   A. Muscle Types (pp. 154–157)
      1. Skeletal Muscle
      2. Smooth Muscle
      3. Cardiac Muscle
   B. Muscle Functions (pp. 157–158)

II. MICROSCOPIC ANATOMY OF SKELETAL MUSCLE (pp. 158–160)

III. SKELETAL MUSCLE ACTIVITY (pp. 160–169)
   A. Stimulation and Contraction of Single Skeletal Muscle Cells (pp. 160–164)
      1. The Nerve Stimulus and the Action Potential
      2. Mechanism of Muscle Contraction: The Sliding Filament Theory
   B. Contraction of a Skeletal Muscle as a Whole (pp. 164–169)
      1. Graded Responses
         a. Muscle Response to Increasingly Rapid Stimulation
         b. Muscle Response to Stronger Stimuli
      2. Providing Energy for Muscle Contraction
      3. Muscle Fatigue and Oxygen Debt
      4. Types of Muscle Contractions—Isotonic and Isometric
      5. Muscle Tone
      6. Effect of Exercise on Muscles

IV. MOVEMENTS, TYPES, AND NAMES OF MUSCLES (pp. 170–174)
   A. Types of Body Movements (pp. 170–173)
   B. Types of Muscles (p. 173)
   C. Naming Skeletal Muscles (pp. 173–174)

V. GROSS ANATOMY OF SKELETAL MUSCLES (pp. 174–182)
   A. Head Muscles (pp. 174–175)
      1. Facial Muscles
         a. Frontalis
         b. Orbicularis Oculi
         c. Orbicularis Oris
         d. Buccinator
         e. Zygomaticus
      2. Chewing Muscles
         a. Masseter
         b. Temporalis
   B. Trunk and Neck Muscles (pp. 175–178)
      1. Anterior Muscles
         a. Sternocleidomastoid
         b. Pectoralis Major
         c. Intercostal Muscles
         d. Muscles of the Abdominal Girdle

        2. Posterior Muscles
            a. Trapezius
            b. Latissimus Dorsi
            c. Erector Spinae
            d. Deltoid
    C. Muscles of the Upper Limb (pp. 178–179)
        1. Muscles of the Humerus That Act on the Forearm
            a. Biceps Brachii
            b. Triceps Brachii
    D. Muscles of the Lower Limb (pp. 179–182)
        1. Muscles Causing Movement at the Hip Joint
            a. Iliopsoas
            b. Adductor Muscles
            c. Gluteus Maximus
            d. Gluteus Medius
        2. Muscles Causing Movement at the Knee Joint
            a. Sartorius
            b. Quadriceps Group
            c. Hamstring Group
        3. Muscles Causing Movement at the Ankle and Foot
            a. Tibialis Anterior
            b. Peroneus Muscles
            c. Gastrocnemius

    VI. DEVELOPMENTAL ASPECTS OF THE MUSCULAR SYSTEM (p. 183)

LECTURE HINTS

1. Students often are confused by the terms *myofiber*, *myofibril*, and *myofilament*. Point out that the longer word represents the smallest structure. Use as many visuals as you can while introducing the ultrastructure of a skeletal muscle cell. Be careful, however, that students understand the three-dimensional nature of this system. A large salted pretzel stick can represent the thick myofilament (if you use your imagination).

2. Wear comfortable clothing, stand on a platform, and demonstrate the body movements to the students. Then, ask them to identify movements as you make them.

3. Instead of dictating the action, origin, and insertion of every muscle you wish students to learn, select five or ten of the muscles and, using diagrams, instruct students on the proper way to study them. Instruct students not to memorize information unless they first understand it. For example, you might want to tell students to find a good diagram of the muscle and note the origin and the insertion. With the picture in view, students might be able to predict the action. Once the origin, insertion and action are understandable, students could make flash cards for convenient, frequent review.

4. It is easy to make your own model to demonstrate sliding filaments. A craft store will have disk-shaped styrofoam pieces that can be used as Z lines on pickup sticks that represent thin myofilaments. Thick myofilaments can be represented by pickup sticks attached to a stand, such as a small ring stand.

5. Discuss disorders of the neuromuscular junction, such as the effects of botulism, snake venom, and myasthenia gravis.

6. Students will be very interested in the effect of various exercise programs on aerobic capacity and muscle mass.

7. If time permits, discuss and demonstrate proper body mechanics.

ANSWERS TO TEXTBOOK REVIEW QUESTIONS
(Questions appear on pp. 193–194)

1. To contract or shorten. When it does so, it causes movement.

2. Skeletal muscle: Long, cylindrical, banded (striated), multinucleate cells; attached to bones and crossing joints; forms the "flesh" of the body and is responsible for all voluntary movement.

   Cardiac muscle: Branching, striated cells containing a single nucleus; interdigitate with one another at tight junctions called intercalated disks; found only in the heart, arranged in spiral bundles; contraction of the heart propels blood into the blood vessels.

   Smooth muscle: Fusiform, uninucleate cells; generally found in cell layers (or sheets) arranged at right angles to one another (one running longitudinally and the other circularly) within the walls of hollow organs; causes substances to move through internal body tracts (digestive, urinary, reproductive, and the like).

3. The arrangement of myofilaments (actin and myosin) within the myofibrils gives each myofilament a banded appearance. The myofibrils are precisely aligned in relation to that banding pattern.

4. They protect, reinforce, and strengthen this delicate muscle tissue. Endomysium, perimysium, and epimysium.

5. Tendons attach muscle to bone. A tendon and an aponeurosis are both tough collagenous structures; they differ in their shape (tendons are cords, whereas aponeuroses are sheetlike).

6. Myoneural junction: The junction of a motor neuron's axonal ending and the sarcolemma of a muscle cell.

   Motor unit: One motor neuron and all the muscle cells it stimulates.

   Tetanus: The tense, contracted state of a muscle.

   Graded response: Different degrees of contraction in response to different levels of stimulation (changes in both the rate and strength of stimuli).

   Muscle fatigue: The inability of a muscle to contract even though it is still being stimulated; usually a result of a lack of oxygen and the accumulation of acids in the muscle tissue.

   Neurotransmitter: A chemical substance released by a neuron when the nerve impulse reaches its axonal endings.

   Aerobic respiration: Metabolic pathways that use O_2 to generate ATP.

   Anaerobic respiration: Metabolic pathway that does not use O_2 to generate ATP.

7. A fatigued muscle has an accumulation of lactic acid (thus a low pH) resulting from a lack of oxygen. A rested or nonfatigued muscle has adequate oxygen delivery; hence, lactic acid does not accumulate.

8. Acetylcholine is released; it diffuses through the synaptic cleft and attaches to receptors on the sarcolemma; sarcolemma permeability to sodium ions increases briefly; sodium ions rush into the muscle cell, changing the electrical conditions of the resting sarcolemma; action potential is initiated and sweeps over the entire sarcolemma; calcium ions are released from storage areas inside the muscle cell; attachment of calcium ions to the myofilaments triggers the sliding of the myofilaments; contraction occurs.

9. Isotonic contractions: Muscle tension remains the same, and the muscle shortens.

    Isometric contractions: Muscle tension increases, and the muscle does not shorten.

10. Muscle tone is a state of continuous, partial contraction of muscles resulting from discontinuous but systematic stimulation by the nervous system. A muscle without tone is paralyzed (unable to contract) and becomes flaccid.

11. Origin: Immovable (or less movable) end.

    Insertion: Movable end; when contraction occurs, the insertion moves toward the origin.

12. Flexion, extension, abduction, adduction, rotation, circumduction, pronation, supination, inversion, eversion, dorsiflexion, plantar flexion. (See pp. 171–172 for illustrations.)

13. A prime mover is a muscle that has major responsibility for causing a particular movement; for example, the gastrocnemius is the prime mover of plantar flexion. Synergist muscles aid prime movers by causing the same movement (but less effectively) or by stabilizing joints or bones over which the prime mover acts; for example, the peroneus muscles (which promote plantar flexion) are synergists of the gastrocnemius muscle. The tibialis anterior muscle causes dorsiflexion of the foot; thus, the gastrocnemius (prime mover for plantar flexion) is its antagonist.

14. Direction of the muscle fibers, for example, rectus abdominis. Relative size of the muscle, for example, gluteus maximus. Location of the muscle with reference to bone, for example, temporalis frontalis. Number of origins of the muscle, for example, triceps brachii. Location of the muscle's origin or insertion, for example, sternocleidomastoid. Muscle shape, for example, deltoid. Action of the muscle, for example, extensor carpi radialis (wrist extensor muscle associated with the radius).

15. Masseter. See Table 6.3, p. 185.

16. Trapezius.

17. Anteriorly, the pectoralis major. Posteriorly, the latissimus dorsi.

18. Prime mover: Biceps brachii.

    Antagonist: Triceps brachii.

19. The four muscles (or muscle pairs) are arranged so their fibers run in different directions, much as sheets of different wood grains are compressed together to make plywood. Like plywood, the abdominal wall musculature is extremely strong for its thickness; it is well constructed for its function as an abdominal girdle.

20. Hamstrings: Extend hip and flex knee.

    Quadriceps: Flex hip (rectus femoris only) and extend knee.

21. Gastrocnemius.

    Plantar flexion.

22. Muscles that are exercised regularly are healthy (with increased endurance), firm and free of superficial fat, and perhaps larger in size (depending on the type of exercise). Resistance-type exercises, such as weight lifting, cause muscles to hypertrophy to meet the increased demands placed on them. Muscles that are not used will atrophy (lose mass) and become weak.

23. With aging, skeletal muscle tissue mass decreases and the relative amount of connective tissue in the muscles increases, causing the muscles to become sinewy. As the muscles decrease in mass, they also decrease in strength. Loss in muscle mass may be partially prevented by regular exercise.

ANSWERS TO AT THE CLINIC QUESTIONS

1. Deltoid, gluteus maximus and medius, vastus lateralis. The vastus is used more often for babies because their arm and hip muscles are poorly developed.

2. Infertility, liver damage (possible cancer), coronary heart disease, mental problems.

3. He ruptured his Achilles tendon, which attaches the gastrocnemius to the heel bone. This accounts for the gap between the calf and the heel, as well as the inability to plantar flex the foot.

4. Any muscle that inserts on the clavicle-trapezius.

5. Eric's oxygen intake has not been adequate to keep his muscles supplied with the oxygen they needed to support prolonged aerobic activity. His heavy breathing will supply oxygen to repay the oxygen debt. His muscle cells were relying on aerobic metabolism, and their oxygen consumption led to breathlessness. When the oxygen ran out, anaerobic metabolism took place, leading to lactic acid accumulation, muscle fatigue, and muscle soreness.

CLASSROOM DEMONSTRATIONS AND STUDENT ACTIVITIES

Classroom Demonstrations

1. Film(s) or other AV materials of choice.

2. Demonstrate muscle contraction (twitch contractions, summation, and tetanus) using a simple myograph or kymograph apparatus and the gastrocnemius muscle of a frog. (Alternatively, a film showing these events might be used. It is important that students be able to visualize these contractile events.)

3. Use models that compare the three types of muscle tissue; point out the unique structural characteristics of each type.

4. Set up a microscope with a slide of a motor unit for class viewing.

5. Use a 3-D model showing the skeletal muscles to point out the muscles students must learn and to demonstrate the proper sites for administering IM injections.

6. Using chicken as an example, discuss the differences between "red meat" and "white meat."

Student Activities

1. Call on students to illustrate the various types of body movements occurring at specific joints: abduction, rotation, flexion, etc. For example, flex your knee, abduct your arm, rotate your head, etc. In a small class, you could play a game, such as Simon Says.

2. Have students work in pairs as follows: One attempts to contract a particular muscle, while the partner provides resistance to prevent that movement. In this way, the muscle will produce its maximal "bulge." Muscles being examined should be palpated in both the relaxed and contracted states by each student. For example, the "demonstrator" can attempt to flex his or her elbow, while the person providing the resistance holds the forearm to prevent its movement. The biceps brachii on the anterior arm will bulge and be easily palpated.

3. Throw out the following descriptions of situations as starting points:

   a. For a class discussion of the importance of skeletal muscles as "heat generators": Suppose you are lost in the woods on a snowy day in sub-zero temperatures. Your tendency is to curl up in the shelter provided by some fallen branches, but common sense tells you to keep moving. What should you do and why?

b. For a discussion of the importance of activity to muscle health: Johnny has just had his leg cast removed after a long period of immobilization. His newly liberated leg looks skinny and twiglike compared to the noncasted leg. Explain.

4. Student assignment for class discussion:

a. Answer the question, "Why do athletes generally perform better if they are allowed to warm up before a competitive event?"

b. Visit a local gym frequented by bodybuilders. Obtain information on the procedures used to build muscle mass and an explanation of how these procedures accomplish that goal.

5. Call out an action, and ask students to provide the name of the muscle responsible for that action. Or, provide the name of a muscle, and ask students to identify antagonists or synergists.

AUDIOVISUAL AIDS

Videotapes

1. *Anatomy of the Musclar System* (IM, 58 min., 1988)
2. *Human Musculature* (BC, 23 min., C, 1989)
3. *The Human Body: Muscular System* (CDR, 23 min., 1980)
4. *Moving Parts* (FHS, VHS or videodisc, 26 min., 1990)
5. *Muscle Fiber* (NDM, 23 min., 1986)
6. *Muscle Power* (FHS, QB-829, VHS or videodisc, 1990)
7. *Muscles* (FHS, 29 min., 1995)
8. *The Physiology of Exercise* (FHS, 15 min., 1988)

Videodiscs

1. *Muscles: Their Structure and Function* (EBF)
2. *Nervous System, Muscular and Skeletal Systems* (NSG)

7 The Nervous System

LECTURE HINTS

1. Because the nervous system is anatomically complex, invest some time at the beginning to make sure students understand how the nervous system is organized and how information flows through the system.

2. Clearly differentiate between the terms nerve and tract, and nucleus and ganglion.

3. The amount of time it will take to cover neurophysiology will depend on the amount of detail you want to provide. You might want to review the movement of ions during a nerve impulse and ask students to predict the change in membrane potential by drawing their own graphs. Or, after discussing the concepts of resting potential and action potential, allow students to place events in their proper order using cards on the board.

4. If time permits, discuss the actions/effects of opiates, antidepressants, and anesthetics or the biology of drug addiction.

5. Students are always interested in the brain and will want to spend as much time as they can learning about it. When discussing the central nervous system, use as many visuals as you can. Prior to class, have students construct a chart in which they list each area of the brain, its location, its structure, and its function. Provide a format for the chart, if possible. In class discuss the information that they have placed in their chart, and if you have time, discuss current brain research and disorders of the brain. Students are particularly interested in the following topics: cerebral palsy, Alzheimer's disease, Parkinson's disease, Huntington's disease, multiple sclerosis, coma, Reye's syndrome, brain tumors, prefrontal lobotomy, hydrocephalus, headaches, the effects of alcohol on the brain, seizure disorders, learning, memory, and differences between the sexes. Collect some student questions in advance and prepare answers.

6. Emphasize the role of the hypothalamus and the medulla in controlling homeostasis, as these brain areas will be important later on in the course.

7. Discuss the variety of central nervous system disorders that occur during the course of HIV infection, resulting not only from opportunistic infections but from direct HIV infection of the CNS.

8. Once students know that the sympathetic branch of the autonomic nervous system mobilizes the body to deal with stress and that the parasympathetic division conserves energy during rest and repose, allow them to make their own chart in which they predict the effects of each division on selected target organs. This is usually easy for students when they think about their own reactions to stress.

ANSWERS TO TEXTBOOK REVIEW QUESTIONS
(Questions appear on pp. 239–240)

1. Nervous system and endocrine system.

2. Anatomical classification: Central nervous system and peripheral nervous system.

   Functional classification: Voluntary division and autonomic division.

3. CNS: brain, spinal cord.

   PNS: Cranial and spinal nerves, ganglia.

4. Neurons are the "nervous cells"; they exhibit irritability and conductivity. The major functions of the glia are protection, support, myelination, and a nutritive/metabolic function relative to the neurons. Schwann cells are myelinating cells.

5. Based on the general direction of the impulse, that is, toward (afferent) or away from (efferent) the central nervous system, and whether or not the neuron is a connecting neuron (interneuron) in the afferent/efferent pathways.

6. A threshold stimulus causes a change in membrane permeability that allows Na^+ to enter the neuron through sodium gates. This causes local depolarization and generates the action potential, which is then self-propagating. This event is quickly followed by a second permeability change that restricts Na^+ entry but allows K^+ to leave the neuron, causing repolarization or resumption of the polarized state. One-way conduction occurs at synapses because axons (not dendrites) release the neurotransmitter.

7. Pain receptors; Pacinian corpuscles (deep pressure) and Meissner's corpuscles (light pressure); temperature receptors (e.g., Krause's end-bulbs [cold] and Ruffini's corpuscles [heat]). The pain receptors are most numerous because pain indicates actual or possible tissue damage.

8. A reflex arc is a neural pathway over which a reflex occurs. The minimum number of components is four (a receptor, an afferent neuron, an efferent neuron, and an effector).

9. Student drawings and responses can be checked by referring to Figure 7.11, p. 207.

10. The pons also has important nuclei that participate in the control of respiratory rhythm.

    The medulla is vital because it contains the major respiratory centers, the vasomotor center (which controls blood vessel diameter, hence blood pressure), and the cardiac centers. Without breathing and heart activity, life stops.

11. The thalamus is a relay station for sensory impulses ascending to the cerebral cortex for interpretation; as impulses pass through the thalamus, one crudely senses that the incoming stimulus is pleasant or unpleasant. The hypothalamus is a major autonomic clearing center whose important functions include body temperature regulation, body water balance, and metabolic control; it also serves as an important center for emotions and drives (sex, rage, pleasure, satiety/appetite, thirst). The cerebellum coordinates skeletal muscle activity (initiated by the cerebrum) and controls balance and equilibrium in accordance with signals received from the inner ear, the eyes, and the proprioceptors of the muscles and tendons.

12. Bone: Enclosed by the skull.

    Membranes: The meningeal membranes—dura mater, arachnoid, and pia mater—enclose the brain within the skull and provide a passage for the circulation of CSF and its return to the blood.

    Fluid: Cerebrospinal fluid (CSF) cushions the brain from physical trauma.

    Capillaries: The capillaries of the brain are permeable only to glucose, a few amino acids, and respiratory gases. Hence, they protect the brain from possibly harmful substances in the blood.

13. Gray matter is neural tissue composed primarily of nerve cell bodies and unmyelinated fibers. White matter is composed primarily of myelinated fibers. In the cerebral hemispheres, the gray matter is outermost (superficial), and the white matter is deep; in the spinal cord, the white matter is outermost and the gray matter is internal or deep.

14. Major reflex center; pathway for ascending sensory impulses and descending motor impulses.

15. Twelve pairs. Purely sensory: Olfactory (I), optic (II), and vestibulocochlear (VIII). Serve external eye muscles: Oculomotor (III), trochlear (IV), and abducens (VI). Regulates heartbeat, etc.: Vagus (X).

16. The head and neck region.

17. Thirty-one pairs. They arise from the dorsal (sensory) and ventral (motor) roots of the spinal cord.

18. Dorsal rami: Posterior body trunk.

    Ventral rami: Limbs and anterior, lateral body trunk.

    (The phrenic nerves of the cervical plexus serve the diaphragm, important in breathing.)

19. Cervical plexus: Diaphragm, shoulder and neck muscles.

    Brachial plexus: Arm.

    Lumbar plexus: Lower abdomen, buttocks, anterior thigh, anteromedial leg.

    Sacral plexus: Lower posterior trunk, posterior thigh and leg.

20. The autonomic nervous system has a chain of two motor neurons (rather than one) extending from the CNS and is controlled involuntarily (rather than voluntarily).

21. The parasympathetic division of the ANS is the "housekeeping system"; it acts to conserve body energy and to keep the body running at minimum levels of energy use during nonemergency periods. Its effect is seen primarily in the normal operation of the digestive system and the urinary system. The sympathetic division is the "fight-or-flight" system; it acts during periods of short-term stress to increase heart rate and blood pressure and to shunt blood glucose levels. Generally, sympathetic activity inhibits digestive system functioning.

22. Although both the sympathetic and presympathetic preganglionic fibers release acetylcholine, their postganglionic fibers (in close contact with the effector organs) release different neurotransmitters. The sympathetic fibers release norepinephrine and the parasympathetic fibers release acetylcholine. These different neurotransmitters produce opposing effects in the effector organs.

23. A concussion occurs when brain injury is slight and the symptoms are mild and transient. Contusions occur when marked tissue destruction takes place. Unconsciousness results because the RAS is injured.

24. Both CVAs and TIAs result from restricted blood flow to brain tissue. CVAs result in permanent or long-lasting deficits, including paralysis, aphasias, and visual disturbances. In TIAs, the disturbances, though similar, are temporary because neurons do not die.

25. Senility is age-related mental deterioration (i.e., changes in intellect, memory, etc.). Permanent causes include factors that deprive neurons of adequate oxygen (such as arteriosclerosis) and degenerative structural changes (as in Alzheimer's disease). Reversible causes include drug effects, low blood pressure, poor nutrition, and hormone imbalances.

ANSWERS TO AT THE CLINIC QUESTIONS

1. Alzheimer's disease.

2. Hypoglossal (XII).

3. The parasympathetic division is involved in the activation of the digestive viscera and with conserving body energy. Following a meal, this system promotes digestive activity and lowers the heart rate and the respiratory rate. The sympathetic division is only minimally active at this time. Therefore, the person will feel "very sleepy." If the person is overweight, he probably should not overexert himself. However, doing the dishes would not be hazardous to his health.

4. Intracranial hemorrhage.

5. Broca's area and adjacent regions of the motor cortex.

6. Cerebral palsy—it will not get worse.

7. Brachial plexus.

8. Schwann cells and oligodendrocytes deposit a fatty coat called myelin around axons. Like the rubber coat around household wires, myelin acts as an electrical insulator.

CLASSROOM DEMONSTRATIONS AND STUDENT ACTIVITIES

Classroom Demonstrations

1. Film(s) or other AV materials of choice.

2. Use a 3-D model (or a transparency) of a motor neuron to point out structural characteristics of nerve cells.

3. Use a dissectible human torso model to show the protected position of the brain and spinal cord within the dorsal body cavity.

4. Use a dissectible human brain model to point out its various structural and functional areas.

5. Use a spinal cord model to illustrate the way the spinal nerves originate from the dorsal and ventral roots and then split into the rami.

6. Using the sciatic nerve of a frog, a stimulator, and an oscilloscope, demonstrate an action potential. This demonstration is available on film.

7. Invite a pharmacologist to discuss the effects of selected drugs on the brain and the entire nervous system.

Student Activities

1. Obtain cow spinal cord sections. Have students make very thin cross-sectional sections with single-edge razor blades and then examine the sections under a dissecting scope or magnifying glass to view the general shape and distribution of the white and gray matter.

2. Obtain a sheep brain with the meninges intact. Have the students examine the brain, paying attention to the toughness of the dura mater, the convoluted nature of the cerebrum, and the location of the various brain stem regions and the cerebellum. Preserved human brains are also available from many vendors.

3. To illustrate the importance of proprioceptors, have a student volunteer hold one hand in back of him/her. Slip a "noose" made of string around that wrist and, by pulling on the string, carry the limb up, down, from side to side, etc. Have the student report the relative position and orientation of the limb after each movement. (Since the string is only around the wrist, touch/pressure will be felt only at that site and the proprioceptors of the joints, muscles, and tendons will be providing the bulk of the sensory input.)

4. Provide reflex hammers and the instructions for producing the patellar and Achilles stretch reflexes and the plantar reflex. Have students work in pairs to produce and observe these reflexes.

5. Without giving any warning, blow a very shrill whistle to produce a startle response in the members of the class. Then, initiate a class discussion on the effects of the sympathetic division of the autonomic nervous system by asking students to indicate which of their body organs were affected by the shrill, unexpected sound and what the organ response was.

6. Have students practice conducting cranial nerve tests. (See Table 7.1 pp. 220–221.)

7. Student assignment for class discussion:

a. Ask students to find an article in a popular magazine that mentions neurotransmitters, and be prepared to tell the class what they have learned. (More than likely, articles will concern the effect of drugs, such as cocaine, on neurotransmitters of the brain, which will provide a good opportunity to discuss the dangers of illegal drug use.)

b. Have students research the effect of alcohol on the brain.

c. Ask students to find an article in a popular magazine that discusses research about sex differences in the brain.

AUDIOVISUAL AIDS

Videotapes

1. *The Addicted Brain* (FHS, 26 min., 1989)
2. *Advancements in Neurobiology and Neurosurgery* (FHS, 22 min., 1995)
3. *The Brain* (IM, 29 min., 1990)
4. *The Brain* (FHS, VHS or videodisc, 20 min., 1995)
5. *The Brain* (EBF, 50 min., 1993)
6. *Decision* (FHS, 26 min., 1989)
7. *Dissection and Anatomy of the Brain* (ALT, 30 min., 1989)
8. *Dream Voyage* (FHS, VHS or videodisc, 26 min., 1989)
9. *The Electric Ape* (FHS, 58 min., 1995)
10. *Exploring the Brain: The Newest Frontier* (HRM, 30 min., 1991)
11. *Human Nervous System* (BC, 29 min., 1992)
12. *Inside Information: The Brain and How It Works* (FHS, 58 min., 1992)
13. *Just Like You and Me: A Video About Epilepsy* (PBS, 20 min., 1990)
14. *Memory: Fabric of the Mind* (FHS, 28 min., 1989)
15. *The Nature of the Nerve Impulse* (FHS, 15 min., 1988)
16. *The Nature of Memory* (FHS, 26 min., 1990)
17. *The Nerve Impulse* (EBF, 21 min., 1971)
18. *Nerves* (IM, 24 min., 1992)
19. *Nerves at Work* (FHS, VHS or videodisc, 26 min., 1990)
20. *The Neuroanatomy Series* (TFI, 1990)
21. *The Neurobiology Series I* (HSC, series of eleven 15 min. films, 1984)
22. *Our Talented Brain* (FHS, VHS or videodisc, 26 min., 1989)
23. *The Peripheral Nervous System* (IFB, 19 min., 1977)
24. *The Power of Addiction* (FHS, 19 min., 1989)
25. *The Sexual Brain* (FHS, 28 min., 1989)
26. *Split Brain and Conscious Experience* (MTI, 18 min., 1982)
27. *The Study of Memory* (FHS, 74 min., 1995)
28. *Within the Human Brain: A Dissection by Dr. Marion C. Diamond* (UCEC, 49 min., 1996)

Videodisc

1. *The Human Brain, The Nervous System* (EBF)

8   Special Senses

LECTURE HINTS

1. Use as many visuals (posters, models, overheads) as you can to teach eye and ear structure.

2. Explain that the eye and the ear are receptors for light and sound impulses. They are also energy transducers because they convert light and sound energy into the electrical energy of an action potential.

3. Point out the importance of other sensations (especially smell) on the perception of taste.

ANSWERS TO TEXTBOOK REVIEW QUESTIONS
(Questions appear on pp. 264–265)

1. Lacrimal glands: Saline solution and lysozyme.

    Meibomian and ciliary glands: Oil.

    Conjunctiva: Mucus.

2. The lacrimal secretions drain via the nasolacrimal duct into the nasal passages.

3. Internal structures: See Figure 8.3 (p. 245).

   Functions: Lens: Major focusing apparatus of the eye. Retina: Photoreceptive (light-sensitive) layer. Choroid: The nutritive (vascular) coat of the eye. Ciliary body: Smooth muscle structure that regulates the shape of the lens for focusing. Iris: Smooth muscle structure that controls the amount of light entering the eye. Sclera: Protective (fibrous) coat. Cornea: Transparent part of the sclera that allows light to enter the eye.

4. Superior, inferior, lateral, and medial rectus muscles; and the superior and inferior oblique muscles.

5. Vitreous humor: Posterior to lens; reinforces eyeball.

   Aqueous humor: Anterior to lens; reinforces eyeball and provides nutrients to the avascular lens and cornea.

6. The portion of the retina with no photoreceptors because it is the site at which the optic nerve leaves the eyeball. Light focused on this spot is not converted into vision or photoreception.

7. The rods focus best in dim light and in peripheral vision; they allow only gray-toned vision. The cones are the color receptors; they function only in bright light and allow for color vision, which is acute discriminatory vision.

8. A small pit in the retinal layer that contains cones only is located lateral to the optic disk in each eye. Anything that must be viewed critically (discriminative vision) is focused on the fovea.

9. Cornea to aqueous humor, through pupil to aqueous humor to lens to vitreous humor to retina (ganglion cell layer to bipolar layer to rods and cones).

10. Photoreceptors (rods and cones) to bipolar cells to ganglion cells to optic nerve to optic tract to thalamus to visual cortex via the optic radiation.

11. The right optic tract contains retinal fibers from the lateral side of the right eye and medical side of the left eye. The right optic nerve contains fibers from both sides of the right eye only.

12. Refraction: Bending of light.

    Refractory structures: Lens, humors, cornea. The lens is the major refractory structure since its refractory action can be controlled.

13. Hyperopia: Farsightedness. The individual has no problem with distance vision, which requires no accommodation; in near vision, however, the image is focused behind the retina because of a lazy lens or foreshortened eyeball.

    Myopia: Nearsightedness. The individual has no problem with near vision, but in distance vision, the image is focused anterior to the retina because of an overly strong lens or elongated eyeball.

    Emmetropia: Normal vision. The image is focused properly on the retina in both near and far vision.

14. As one ages, the lens loses its elasticity, causing it to focus less acutely for close vision. Hyperopia.

15. When two (or more) different cone types are stimulated simultaneously the color perceived is intermediate (in wavelength or light) between them. Thus, stimulation of red and green cones produces the sensation of yellow; when all cones are excited, we perceive white (all colors = white).

16. The ophthalmoscopic examination allows the examiner to determine the condition of the retina. Many blood vessel problems (e.g., high blood pressure and diabetes mellitus) can be detected early from such an examination; hence, it is extremely important in diagnosis.

17. Close vision involves both convergence and accommodation. Convergence requires the activation of the external eye muscles (primarily the medial recti) and the ciliary body within the eye. Distance vision requires none of these specific muscle actions.

18. The pupillary response (i.e., pupil constriction in response to light) is very important in preventing damage to the eye's photoreceptors from intense light.

19. Outer ear: The auricle (pinna) has no function in humans. The external auditory canal conducts sound vibrations from the external environment to the eardrum (tympanic membrane), which is set into sympathetic vibration. Involved in hearing only.

    Middle ear: The ossicles, which span the tympanic cavity, conduct vibrations received from the tympanic membrane to the fluids of the inner ear. The auditory tube allows middle ear pressures to be equalized with those of the atmosphere. Involved in hearing only.

    Inner ear: Semicircular canals, vestibule, cochlea. The semicircular canals and vestibule function in equilibrium maintenance, whereas the cochlea functions in hearing.

20. Eardrum to ossicles (hammer to anvil to stapes) to oval window to perilymph in the bony cochlea to endolymph in cochlear duct to basilar membrane and organ of Corti.

21. Sensorineural deafness arises from damage to neural structures (cochlear nerve or auditory cortex cells) due to stroke or other trauma. Conductive deafness arises from any interference with the conduction of vibrations from the outer to the inner ear due to ear wax accumulations, otitis media, fusion of the ossicles, or pressure imbalance between the middle and outer ear.

22. Static equilibrium has to do with the ability to determine the at-rest position of the head in space (e.g., up or down). Dynamic equilibrium relates to the angular/rotatory movements of the body in space.

23. Inner ear vestibular apparatus (semicircular canals and vestibule), eyes, proprioceptors.

24. Gustatory receptors. In the taste buds—primarily in papillae on the tongue, but also in the buccal mucosa and on the palate and pharynx mucosae.

25. Sweet, sour, bitter, salty.

26. In the superiormost mucosa of the olfactory passageways. This is a poor site functionally because most air (which carries odors) passes inferior to this site.

27. Chemoreceptors. Both receptors respond to chemicals in aqueous solution.

28. Presbyopia (basically age-related farsightedness), presbycusis (sensineural deafness due to cumulative hearing insults), and dulled sense of taste and smell (these receptors are replaced more slowly).

29. Of all the special senses, vision requires the most "learning." The eye must first enlarge, then external muscle coordination must develop, as must focusing ability and color vision. Depth perception develops relatively late, maturing just before school age.

ANSWERS TO AT THE CLINIC QUESTIONS

1. His hearing receptors in the organ of Corti are being damaged (and are dying) by the excessively loud noise; sensineural deafness is occurring.

2. The children probably had conjunctivitis, caused by bacteria or viruses; it is highly contagious.

3. High intraocular pressure generally indicates that the patient has glaucoma.

4. The left optic tract carries all the visual information from the right half of the visual field. Thus, Lionel cannot see the right half of visual space.

5. (a) The surgeon was trying to remove the tiny stapes from the oval window, to which it had rigidly fused, causing conduction deafness. (b) Such delicate surgery is difficult because the stapes is so tiny. Here, it proved impossible to remove this ossicle neatly enough to clear the oval window.

6. Myopia refers to nearsightedness. Parallel rays from objects far away are focused in front of her retina. She will need a concave corrective lens to diverge these rays so that they will converge on the retina.

CLASSROOM DEMONSTRATIONS AND STUDENT ACTIVITIES

Classroom Demonstrations

1. Film(s) or other AV materials of choice.

2. Use 3-D models of the eye and ear to demonstrate the structural characteristics of these special sense organs.

3. Dissect a cow's eye so that students can observe the internal eye structures (iris, ciliary body, lens, vitreous humor, retina, choroid). Point out how fragile the retina is, and call the students' attention to the fact that it is attached at a single point (optic disk).

4. Spray a cologne with a light, citrus-based or "woodsy" scent into the air and ask students to indicate by raising their hands when they first smell it. Then go on to other matters and about 5 minutes later, ask how many can still smell the cologne. Use this as a jumping-off point for your discussion on adaptation of the sensory apparatus.

Student Activities

1. To examine the important connection between smell and taste in the "taste" and identification of foods, select one student to be the subject in the following experiment. The student closes his or her eyes and pinches the nostrils closed. Then he or she should attempt to distinguish mozzarella cheese from hard-boiled egg white by taste alone. If the attempt is unsuccessful, the nostrils should be released and the test conducted again. (Note: The cheese and egg white should be cut into pieces of similar shape and size, perhaps small cubes. A sip of water should be taken between samplings.)

2. Obtain Ishihara's color plates, and have students test themselves for evidence of color blindness.

3. Student assignment for class discussion: Visit a local nursing home and volunteer services for an afternoon. During this interval, observe and record all examples of diminished capabilities of the special senses (vision, smell, taste, equilibrium, etc.) observed in the elderly clients. Be prepared to discuss how and why these deficits threaten homeostasis.

4. Write the names of various structures of the eye and the ear on cards, and attach them to the board (using magnets or tape). Have students come to the board to arrange the cards in the order that light or sound signals travels through these receptors.

5. Bring an eye chart to class so that students can assess their visual acuity. Have students find their blind spot.

6. Collect examples of optical illusions, and give students the opportunity to explain them.

AUDIOVISUAL AIDS

Videotapes

1. *Cataracts* (PLP, 10 min., 1989)
2. *Ears: Have You Heard the Latest* (CIF, 11 min., 1981)
3. *Eye Dissection and Anatomy* (IM, 16 min., 1989)
4. *Eyes and Ears* (FHS, VHS or videodisc, 26 min., 1990)
5. *The Human Eye* (IFB, 14 min., 1978)
6. *The Senses* (FHS, 20 min., 1995)
7. *The Senses: How We Know* (NGS, 15 min., 1992)
8. *Smell and Taste* (FHS, 26 min., 1990)

Videodisc

1. *The Ears and Hearing, The Eyes and Seeing* (EBF)

9   The Endocrine System

LECTURE HINTS

1. Explain to students that although hormones circulate through the body freely, all tissues are not susceptible to all hormones. In order for a tissue to respond to a particular hormone, its cells must have specific receptors for that hormone. Diseases of the endocrine system can result from a hormonal imbalance or a receptor imbalance.

2. Stress the anatomical and functional relationship between the nervous system and the endocrine system.

3. A discussion of endocrine disorders, such as pituitary dwarfism, gigantism, diabetes mellitus, diabetes insipidus, Addison's disease, cretinism, acromegaly, and hyperthyroidism, will make this lecture more interesting. A discussion or reading about diabetes mellitus and the pioneering work of Banting and Best adds human interest to this topic.

ANSWERS TO TEXTBOOK REVIEW QUESTIONS

(Questions appear on pp. 292–293)

1. a. Nervous system control is extremely rapid (milliseconds to seconds), whereas endocrine control takes minutes to days to bring about its effects.

   b. Nervous system communication is via electric impulses, whereas the endocrine system uses blood-borne chemical "messengers" (hormones).

   c. The nervous system controls short-term processes, such as stimulation of muscle contraction and glandular secretion, whereas the endocrine system controls processes that go on (sometimes continuously) for long periods of time, such as growth and maturation, metabolism, and the functioning of the reproductive system.

2. The mixed endocrine organs are the pancreas and gonads. The purely endocrine glands are the anterior pituitary, thyroid, parathyroid, adrenal, thymus, and pineal gland.

3. Hormone: An amino acid or steroid-based chemical substance produced by endocrine organs and liberated to the blood; activates target organs in specific ways.

4. Hormonal stimulation: Stimulation by TSH and ACTH of the anterior pituitary of the thyroid and adrenal cortex (respectively) to release their hormones.

   Humoral stimulation: Stimulation by high glucose levels to release insulin; stimulation by high blood calcium levels to release calcitonin.

   Nerve stimulation: Stimulation by sympathetic nerve fibers of adrenal medulla to release epinephrine (adrenalin) to the blood.

5. Negative feedback, the process by which the products of a chemical reaction "feedback" and inhibit their further synthesis, regulates the activity of all endocrine glands controlled by anterior-pituitary-tropic hormones, as well as those controlled by blood levels of nonhormonal substances. Most simply, when a hormone is at low levels in the blood, its synthesis is stimulated either by the release of an anterior-pituitary-tropic hormone (e.g., thyroxine production and release are stimulated by TSH) or by changing blood levels of certain substances (e.g., PTH production is stimulated by low blood calcium levels and insulin release is stimulated by high levels of blood glucose). As blood levels of the stimulated hormones increase, the stimulus substance is either turned off (in the case of tropic hormones) or ceases to exist (because hormonal action results in a "correction" of the blood levels of the trigger substances). Once there is no stimulus, the previously stimulated endocrine organ decreases its hormone output, and hormone levels drop once more. The cycle repeats again and again as hormone blood levels increase and decrease via the process of negative feedback.

6. Target cells: The cells that respond to a specific hormone. A hormone's target cells have plasma membrane receptors or internal receptors that are able to bind that particular hormone. Tissue cells that are not target cells of a particular hormone lack those particular receptors.

7. Anterior pituitary: Protrudes from the inferior surface of the brain, encased in the sella turcica of the sphenoid bone. Produces: (a) STH, which causes overall body growth but particularly skeletal and muscular growth; a lack during childhood leads to pituitary dwarfism; over-secretion produces gigantism (children) or acromegaly (adult). (b) Prolactin (PRL), which stimulates lactation in females; an excess leads to inappropriate lactation and impotence. (c) Gonadotropic hormones FSH and LH, which stimulate the production of ova/estrogen/progesterone in females and sperm/testosterone in males; a lack results in sterility. (d) TSH, which stimulates the production of thyroxine by the thy-

roid gland; a lack results in hypothyroidism (cretinism in children, myxedema in adults); hypersecretion produces Graves' disease. (e) ACTH, which regulates the activity of the adrenal cortex; hypersecretion results in Cushing's syndrome; hyposecretion results (secondarily) in Addison's disease. The posterior pituitary releases two hormones made by the hypothalamus, ADH and oxytocin; a lack of ADH leads to diabetes insipidus.

Pineal gland: Found at the superoposterior end of the third ventricle in the brain. The pineal releases melatonin, which inhibits precocious sexual development in humans. Early hypersecretion results in sexual maturity earlier than normal.

Thymus: Found in the anterior thorax, overlying the trachea and heart. Its hormone, thymosin, serves to "program" the T lymphocytes of the immune system for recognition of self from nonself. Athymic individuals lack the ability to mount any type of immune response.

Pancreas: Located in the abdomen, in the mesentery between the stomach and duodenum. Produces: (a) Insulin, basically a hypoglycemic hormone that promotes the uptake and metabolism of glucose by body cells; a lack of insulin leads to diabetes mellitus. (b) Glucagon, basically a hyperglycemic hormone that promotes the release of glucose by the liver when blood glucose levels are low; no documented hypersecretion or hyposecretion problems.

Ovaries: Located in the abdominopelvic cavity, lateral to the uterus. Produce estrogen(s) and progesterone. Lack of these hormones leads to the inability to conceive or bear children.

Testes: Located in the scrotal sac, medial to the superior thighs. Produce testosterone, a lack of which leads to the inability to produce sperm.

(Note: Only those hypersecretion or hyposecretion effects considered problematic have been noted.)

8. The adrenal cortex produces glucocorticoids (e.g., cortisone, hydrocortisone), which are important in aiding the body to resist long-term stressors, such as bodily trauma and anxiety. Glucocorticoids mobilize blood sugar and decrease the severity of the inflammatory response. The adrenal medulla produces epinephrine, which, together with the sympathetic nervous system, aids the body to react quickly to short-term stressors by diverting blood to the heart, brain, and skeletal muscles and by increasing blood sugar, blood pressure, and heart rate.

9. Releasing hormones of the hypothalamus (in addition to feedback inhibition).

10. Tropic hormones are those hormones produced by the anterior pituitary that cause some other endocrine organ to release its hormones.

11. The posterior pituitary is not a true endocrine organ because it does not produce the hormones it liberates. Rather, it stores and releases hormones produced by the hypothalamus.

12. A tumor.

13. Insulin: glucagon, glucocorticoids, and epinephrine.

    PTH: calcitonin.

14. ADH and aldosterone. ADH is produced by the hypothalamus and released by the posterior pituitary; causes the kidney tubules to increase their uptake and retention of water (from the kidney filtrate). Aldosterone, an adrenal cortical hormone, regulates sodium ion (and, secondarily, potassium ion) concentration in the blood. When sodium ion concentration is too low, aldosterone is released and causes the renal tubules to increase their reabsorption of sodium ions (and water follows passively).

15. A goiter is the result of a lack of iodine in the diet. The thyroid hormones thyroxine (T_4) and triiodothyronine (T_3) contain iodine, so when iodine is not present, the functional hormones cannot be made. The anterior pituitary (via TSH) continues to stimulate the thyroid follicles to produce T_3 and T_4 when levels in the blood are low. This continued but ineffective stimulation leads to enlargement of the thyroid gland.

16. As their ovaries cease to function, aging women undergo menopause, experiencing such symptoms as hot flashes, and becoming susceptible to osteoporosis and arteriosclerosis. Both males and females tend to become increasingly hypothyroid with age; they also become more susceptible to adult-onset diabetes as pancreatic function declines. In addition, the decreasing amounts of antistress hormones lower resistance to infectious disease.

ANSWERS TO AT THE CLINIC QUESTIONS

1. The dysfunction is hypersecretion of androgens, possibly from a tumor of the adrenal cortex.

2. The diagnosis is hyposecretion of growth hormone. The prescription is commercial pituitary growth hormone. The reason the girl might reach her growth potential is that the epiphyseal plates of her bones have not yet closed, allowing additional growth of the skeleton and body, in response to the hormone.

3. Pitocin and oxytocin stimulate contraction of the smooth muscle of the myometrium, thus strengthening the uterine contractions to expel the baby.

4. The catecholamines epinephrine and norepinephrine. The usual cause of hypersecretion is a tumor in the adrenal medulla. The catecholamines promote a rise in blood sugar (hyperglycemia), whereas thyroid hormones trigger glucose catabolism.

CLASSROOM DEMONSTRATIONS AND STUDENT ACTIVITIES

Classroom Demonstrations

1. Film(s) or other AV materials of choice.

2. Using a human torso model, indicate the body locations of the major endocrine organs.

3. Obtain photographs of individuals with endocrine disturbances, for example, someone with exopthalamos/Graves' disease, Cushing's syndrome, gigantism, etc. Use these pictures to support the presentation of hyper- and hyposecretion disorders.

4. Demonstrate the effect of epinephrine on heart rate (i.e., short-lived increase in heart rate) using a frog. Double-pith a frog and place it dorsal-side-down on a dissecting pan. Carefully open its ventral body cavity by making a vertical incision with scissors. Identify the beating heart and carefully cut through the saclike pericardium to expose the heart tissue. Have students visually count the heart rate for 1 minute and then record that figure. Flush the heart with epinephrine solution and have students count and record the heart rate for 5 successive minutes. This experiment can be found on videodisc and film (see Audiovisual Aids below).

Student Activities

1. Use a small fish to demonstrate the effects of hyperinsulinism and to initiate a discussion of what measure should be taken with people exhibiting signs of hyperinsulinism.

Prepare two beakers: Beaker A contains 200 mL water and 10 to 15 drops of commercial insulin; beaker B contains 200 mL of 10% glucose solution. Place a small fish (goldfish or sunfish) in beaker A. Have students observe the fish's actions as insulin diffuses into its bloodstream and note how long it takes for the fish to become comatose. When the fish is comatose, transfer it to beaker B, again have students observe and record its actions, and note how long it takes for recovery. After completing the observations, return the fish to the aquarium.

2. Student assignment for class discussion:

a. Read about juvenile (insulin-dependent) and adult-onset (noninsulin-dependent) diabetes and be prepared to discuss why one form absolutely requires insulin for treatment whereas the other can be treated with diet and/or exercise.

b. Research the use of corticosteroids in nonendocrine disorders. Be prepared to describe the advantages and disadvantages of using these drugs.

c. Research the current applications and experimental uses of recombinant human growth hormone.

AUDIOVISUAL AIDS

Videotapes

1. *Animal Hormones I and II: Principles and Functions* (EIL, 60 min., 1987)
2. *The Chemistry of Life: Hormones and the Endocrine System, Parts 1–3* (HRM, 1981)
3. *The Endocrine System* (CIF, 48 min., 1988)
4. *The Endocrine System* (IM, 20 min., 1993)
5. *Glands and Hormones* (IM, 25 min., 1994)
6. *Hormones and the Endocrine System* (IM, 45 min., 1981)
7. *The Living Body: Messengers* (FHS, VHS or videodisc, 27 min., 1990)

Videodisc

1. *The Respiratory System, The Endocrine System* (EBF)

10   Blood

LECTURE HINTS

1. Discuss how the reaction with Wright's stain is the basis for naming the granulocytes.

2. After discussing hemostasis with students, ask them why they think bleeding time (the amount of time it takes for bleeding to stop) is less than clotting time (the amount of time it takes for a clot to form).

3. Explain to students that because the liver is the source of albumen and globulins, liver disease can lead to edema and bleeding disorders.

4. Most students realize that a lack of red blood cells leads to anemia. Many students, however, do not realize that polycythemia can lead to elevated blood viscosity and hypertension. Discuss this as a harmful side effect of blood doping.

5. Students will be very interested in a discussion of anemias, leukemias, and hemophilia. A clarification of classification schemes used for naming anemias and leukemias might be very useful.

6. Present information about the clinical uses of recombinant erythropoietin, colony-stimulating factors, and tissue plasminogen activator.

7. Although there are many blood groups, the ABO and the Rh agglutinogens are clinically the most important. Other blood groups are often analyzed for purposes such as resolving inheritance disputes or in forensics. These other groups can, on rare occasion, be clinical problems.

8. Students usually have some initial difficulty understanding blood groups. It might be helpful to explain that agglutinogens are antigens that are found on RBC surfaces. Agglutinins, on the other hand, are antibodies made by lymphocytes and found in the plasma.

   Constantly remind students that people only make antibody (agglutinin) against antigen (agglutinogen) that they do not have, and that you can safely donate blood only if the recipient already possesses the donated agglutinogen.

9. Students are usually interested in Rh incompatibilities, even though treatment protocols initiated in the past few decades have made erythroblastosis fetalis less common. ABO incompatibilities are more frequent and can also cause symptoms, but they are usually not as severe.

10. If time permits, discuss the genetics of ABO and Rh inheritance. Students will have fun determining the blood types of their parents, siblings, and children in an effort to figure it all out!

ANSWERS TO TEXTBOOK REVIEW QUESTIONS
(Questions appear on p. 312)

1. Average blood volume of an adult is approximately 5.5 L.

2. When blood is well oxygenated, it is bright red in color. Deoxygenated or oxygen-poor blood is dull red-bluish in color.

3. a. Gases (oxygen, carbon dioxide, nitrogen).

   b. Foodstuffs (glucose, lipids, fatty acids, cholesterol, amino acids).

   c. Ions (calcium, iron, chloride and other salts).

   d. Hormones.

   e. Metabolic wastes (urea, uric acid, ammonia, creatinine).

   f. Antibodies.

   g. Clotting proteins.

   h. Other proteins, including various enzymes, albumin, and transport proteins.

4. Formed elements: Blood cells (or living portion) found in blood. They include erythrocytes, leukocytes, and thrombocytes. Erythrocytes are the most numerous; leukocytes and thrombocytes comprise the buffy coat.

5. The average lifespan of a RBC is between 100 and 120 days. Because the RBC has no nucleus, it is unable to restore is proteins (most importantly its enzymes) as they begin to break down. Without the ability to form new proteins, the ability to produce ATP declines, and the cell begins to fragment.

6. Anemia: A decrease in the oxygen-carrying ability of the blood. Possible causes: A lack of dietary iron, hemorrhage, depression or destruction of the bone marrow, a lack of vitamin B12, bacterial infections leading to blood lysis, genetic defects.

7. Granular WBCs: Neutrophils—phagocytes; eosinophils—act during allergy and parasitic infections; basophils—release histamine and heparin during inflammatory reactions.

   Agranular WBCs: Monocytes—phagocytes; lymphocytes—part of the body's immune system; form antibodies and act directly against foreign substances that have managed to invade the body.

8. With a severe infection, leukocytosis occurs; thus, 15,000/mm³ would be expected, since WBCs are important in defending the body against infection.

9. Hemocytoblast. The myeloid stem cell produces erythrocytes, platelets, and all leukocytes except the lymphocytes. The lymphoid stem cell gives rise to lymphocytes only.

10. Hemostasis is initiated by a break in the blood vessel wall (or lining), which causes platelets to cling to the damaged site. Once attached, the platelets release serotonin, which triggers vasoconstriction. Injured tissue cells release thromboplastin, which interacts with platelet phospholipids (PF$_3$), Ca^{2+}, and plasma clotting factors to form prothrombin activator. Prothrombin activator converts prothrombin to thrombin. Thrombin, an enzyme, then converts soluble fibrinogen molecules into long fibrin threads, which form the basis of the clot.

11. The liver is the source of fibrinogen and several other factors that are necessary for clotting.

12. The type of specific agglutinogen or proteins present on the RBC cell membrane. Agglutinins are antibody molecules present in the plasma that agglutinate foreign RBCs (i.e., RBCs with different agglutinogens present on their cell membranes).

13. A, B, AB (least common), and O (most common).

14. Transfusion reaction: The clumping and lysis of foreign (donor) RBCs by the host's agglutinins when a mismatched blood type is infused. The clumping of RBCs may clog capillaries, and the released hemoglobin may block kidney tubules, eventually leading to renal shutdown.

15. The Rh⁻ person does not have preformed antibodies against Rh⁺ blood. However, if such a person once receives Rh⁺ blood, the Rh antigens on the RBC cell membranes are recognized as foreign, and anti-Rh antibodies are formed. On the second (and subsequent) transfusion of Rh⁺ blood, a typical antigen-antibody reaction occurs in which the mismatched donor's RBCs are agglutinated and lysed.

16. High. The hematocrit, or packed cell volume, provides a good estimate of the relative volume of RBCs. RBCs are the site of hemoglobin in the blood, so if their number increases, one would also expect the total amount of hemoglobin in the blood to increase (barring certain problems with hemoglobin formation, such as iron deficiency).

17. Leukemia, anemias, and clotting disorders.

ANSWERS TO AT THE CLINIC QUESTIONS

1. Erythropoietin.

2. Aplastic anemia; short-term: transfusion; long-term: bone marrow transplant; packed red cells.

3. Hemorrhagic anemia (and iron deficiency).

4. Polycythemia vera.

5. He has secondary polycythemia which is the body's attempt to restore the homeostasis of blood oxygen levels in an environment where the air is thinner and less oxygen is available. The RBC count will begin to decline as the body readapts to oxygen levels in "Boston air" because erythropoietin levels decline when excess oxygen is carried in the blood.

6. A crushed or torn artery sustains more tissue damage than a clearly severed one. Since the vascular spasm is proportional to the amount of tissue damage, a crushed or torn artery will lead to less blood loss.

CLASSROOM DEMONSTRATIONS AND STUDENT ACTIVITIES

Classroom Demonstrations

1. Film(s) or other AV materials of choice.

2. Use models of the various blood cells.

3. Display equipment used to perform a hematocrit, sed rate, and blood cell counts. Describe how these tests are conducted and the information they yield. Then run a hematocrit so that students can see the difference in the volume of plasma and formed elements.

4. Set up a prepared blood smear under a demonstration microscope. Show as many of the different WBCs in the field as possible for student viewing.

5. Set up prepared slides or use 35-mm slides showing pathologies such as sickle cell anemia, leukemia, erythroblastosis fetalis, and iron deficiency anemia. Ask students to discuss how these blood smears differ from blood smears from healthy individuals.

6. Invite a speaker from a local blood bank to talk to the class.

Student Activities

1. Provide blood typing sera, and have students type their own blood. (All lancets and disposable items are to be placed immediately into a disposable autoclave bag after use, and used slides are to be placed in a beaker containing freshly prepared 10% bleach solution and soaked for at least 2 hours. Both the autoclave bag and the slides are to be autoclaved for 15 minutes at 121°C and 15 lb pressure to insure sterility. After autoclaving, the autoclave bag may be discarded in any disposal container; the glass slides may be washed with laboratory detergent and prepared for use.)

   Another alternative is to purchase aseptic blood samples from a company such as Carolina Biological Supply Company. Although this will eliminate the need for students to draw their own blood, care and proper handling should be strictly enforced.

2. Provide a sample of centrifuged animal blood so that the students can examine the consistency, texture, and color of plasma. Have pH paper available so they can determine its pH. Use this as a lead-in to a discussion of the composition and importance of plasma.

3. Student assignment for class discussion: Find out which diseases are transmitted via blood, and be ready to explain why many of these diseases are increasing in incidence, and why the careful handling of blood in the clinical agency is vitally important.

4. Collect data from the class about their own blood types. Compare these data to expected frequency.

AUDIOVISUAL AIDS

Videotapes

1. *Blood* (FHS, 20 min., 1995)

2. *Blood Is Life* (FHS, 45 min., 1995)

3. *Blood—River of Life, Mirror of Health* (HRM/EI, 60 min., 1990)

4. *The Blood System: A Liquid of Life* (MFM, 45 min., 1989)

5. *Blood Smear* (HSC, 9 min., 1989)

6. *Blood: The Vital Humor* (FHS, 26 min., 1990)

7. *Blood Group Systems and Blood Bank Serology* (MFM, 35 min., 1982)

8. *Leukemia* (FSH, 22 min., 1995)

9. *The Life of the Red Blood Cell* (KSU, 10 min., 1990)

10. *Life Under Pressure* (FHS, VHS or videodisc, 23 min., 1990)

11. *New Directions in Blood Transfusions* (FHS, 20 min., 1995)

12. *Our Nation's Blood Supply: The Next Threshold for Safe Blood* (FHS, 22 min., 1995)

13. *William Harvey and the Circulation of Blood* (FHS, 22 min., 1995)

11   The Circulatory System

SUGGESTED LECTURE OUTLINE

LECTURE HINTS

1. Show a video of a beating heart, ideally with heart sounds. Stress that while the right side of the heart is a pulmonary pump and the left side a systemic pump, both atria contract at the same time and both ventricles contract at the same time.

2. Although the left side of the heart generates more pressure than the right side, approximately the same volume of blood is ejected from each side per beat. Ask students to think about what would happen if this was not the case. Follow up with a discussion of congestive heart failure.

3. Stress that the only function of the valves of the heart is to ensure the one-way conduction of blood. Explore with students the consequences of incompetent or stenotic valves.

4. Transplanted hearts are often not reinnervated. Starling's law of the heart is very important for maintaining circulation.

5. Explore with students the consequences of atrial and ventricular septal defects.

6. A very basic misconception students have is that arteries carry oxygenated blood and veins carry deoxygenated blood. Remind students that arteries are defined as vessels that carry blood away from the heart and the veins are defined as vessels that carry blood toward the heart.

7. To explain the elasticity of arteries, compare them to a balloon that expands when it fills with air and deflates rapidly and maintains pressure, losing air, when it is let go.

8. Students get confused when the name of a blood vessel changes as it travels into a new area of the body. The left subclavian artery, for example, becomes the left axillary artery, and this then becomes the left brachial artery. Compare these arteries to streets in your community that change their name in a similar fashion.

9. When discussing fetal circulation, remind students that it is very similar to pulmonary circulation in the sense that the placenta serves as a source of oxygen for the fetus. Discuss with students what happens to fetal circulation after birth.

10. Explore factors that elevate or reduce mean arterial pressure using the equation MAP = CO x HR. Give students a situation, and ask them to predict its effect on MAP. Discuss the effects of aging on MAP.

11. Show a picture or a slide of the mediastinal lymph nodes of a smoker.

12. If time permits, discuss hypertension, including prevention and methods used to treat it.

13. Compare lymph capillaries and blood capillaries.

14. Discuss the consequences of lymphedema that occur after breast surgery or as a result of elephantiasis.

15. Discuss the reasons for splenectomy and tonsillectomy.

ANSWERS TO TEXTBOOK REVIEW QUESTIONS
(Questions appear on pp. 352–353)

1. The heart is located in the thorax, slightly left of center. It is flanked laterally by the lungs and its apex rests on the diaphragm at the level of the fifth intercostal space.

2. See Figure 11.2 (p. 317).

3. Right atrium to right ventricle to pulmonary trunk to right and left pulmonary arteries to pulmonary capillaries of the lungs to pulmonary veins to left atrium of the heart. Pulmonary circuit or pulmonary circulation.

4. The systemic circulation provides oxygen-rich blood to all tissues of the body; it is the functional blood supply of all organs. The pulmonary circulation serves only to bring venous blood to the lungs for unloading of carbon dioxide and loading of oxygen, then to return the blood to the left side of the heart.

5. Heart valves enforce a one-way flow of blood through the heart. Yes, the heart can function with leaky valves (just as a water pump can), but it is less efficient.

6. The coronary arteries feed the heart muscle itself (i.e., the myocardium), so if they are blocked by a clot, the heart muscle will not receive adequate glucose and oxygen, and its cells will begin to die.

7. The pericardial (serous) fluid acts as a lubricant to decrease friction as the heart beats.

8. Systole: Period of contraction of the heart (usually refers to ventricular contraction).

   Diastole: Period of relaxation of the heart musculature.

   Stroke volume: The amount of blood pumped out by a ventricle with each contraction.

   Cardiac cycle: The time for one complete heartbeat, from the beginning of one systole to the beginning of the next.

9. Ventricles.

10. The heart has an intrinsic ability to beat. Whereas the nervous system may increase or decrease its rate, the heart continues to beat even if all nervous connections are cut.

11. The intrinsic conduction system enforces a unidirectional depolarization on the heart and causes it to beat as a coordinated unit. Since the rate of depolarization of its pacemaker (the SA node) is more rapid than that of the myocardial muscle, it causes the heart to beat at a more rapid rate than it would otherwise.

    Elements: SA node, AV node, AV bundle, bundle branches, Purkinje fibers.

12. Lub-dup. Closure first of the AV and then of the semilunar valves.

13. Activity of the sympathetic nervous system (as during physical or emotional stress), excess or lack of certain vital ions, increased temperature.

14. Tunica intima: A single layer of squamous epithelium; provides a smooth, friction-reducing lining for the vessel.

    Tunica media: A middle layer, consisting of smooth muscle and connective tissue (primarily elastic fibers). The elastic fibers provide for stretching and then passive recoil of vessels close to the heart, which are subjected to pressure fluctuations; the smooth muscle is activated by the sympathetic nervous system when vasoconstriction (and increases in blood pressure) is desired.

    Tunica externa: The outermost layer, made of fibrous connective tissue; basically a protective and supporting layer.

15. Capillary walls are essentially just the tunica intima (endothelium plus the basement membrane); thus, they are exceedingly thin. This is advantageous because the capillary walls are the site of the exchange of substances between the tissue cells and the blood.

16. Arteries are much closer to the pumping action of the heart and must be able to withstand the pressure fluctuations at such locations. Veins, on the distal side of the capillary beds of the tissues, are essentially low-pressure vessels that need less strength/support/elasticity than do arteries.

17. The presence of valves, the milking action of skeletal muscles against the veins as the muscles contract, the respiratory pump (pressure changes in the thorax during breathing).

18. Pulmonary arteries carry oxygen-poor blood and pulmonary veins carry oxygen-rich blood. Umbilical arteries carry oxygen-poor blood from the fetus and the umbilical vein carries the most oxygen-rich blood to the fetus.

19. Right wrist: Left ventricle to ascending aorta to aortic arch to brachiocephalic artery to subclavian artery to axillary artery to brachial artery to radial (or ulnar) artery to capillary network of wrist to radial (or ulnar) vein to brachial vein to axillary vein to subclavian vein to right brachiocephalic vein to superior vena cava to right atrium of the heart.

    Right foot: Left ventricle to ascending aorta to aortic arch to descending aorta to right common iliac artery to external iliac artery to femoral artery to popliteal artery to anterior tibial artery to dorsalis pedis artery to capillary network to anterior tibial vein to popliteal vein to femoral vein to external iliac vein to common iliac vein to inferior vena cava to right atrium of the heart. (Alternatively, the sequence between the underscored points could be stated as: dorsal venous arch to great saphenous vein.)

20. Internal carotid arteries and vertebral arteries. Circle of Willis.

21. The hepatic portal circulation carries nutrient-rich blood from the digestive viscera to the liver for processing before the blood enters the systemic circulation. A portal circulation involves a capillary bed that is both fed and drained by veins; the usual circulation has a capillary bed that is fed by arteries and drained by veins.

22. In a fetus, both liver and lungs are nonfunctional (the liver relatively so). The ductus venosus bypasses the liver. The ductus arteriosus and the foramen ovale bypass the lungs. The umbilical vein carries nutrient-rich and oxygen-rich blood to the fetus through the umbilical cord.

23. Pulse: The alternate expansion and recoil of an artery that occur with each heartbeat.

24. Wrist: Radial artery.

    Front of the ear: Temporary artery.

    Side of the neck: Carotid artery.

    Groin: Femoral artery.

    Back of knee: Popliteal artery.

25. Blood pressure: Pressure that blood exerts on the walls of the blood vessels.

    Systolic pressure: Pressure exerted by blood on the arterial walls during ventricular contraction.

    Diastolic pressure: Pressure exerted by blood on the arterial walls when the ventricles are relaxing (that is, during diastole).

26. Blood pressure is the force that maintains the circulation of blood through the blood vessels.

27. Cardiac output is increased by increased venous return and increased heart rate. Peripheral resistance is increased by decreased diameter of the blood vessels and increased blood viscosity.

28. Hemorrhage causes blood pressure to drop initially because the decreased blood volume leads to decreased venous return and decreased cardiac output. But as the sympathetic nervous system initiates vasoconstriction (and increases peripheral resistance), blood pressure is increased (within limits). Blood pressure is normally highest in the recumbent position and lowest immediately after standing up; however, the sympathetic nervous system quickly compensates in a healthy individual. Very often an individual can become hypotensive after remaining still in the sitting position for an extended period.

29. Lymphatic vessels: To pick up excess fluid and proteins that have leaked from the blood vascular system and to return them to the blood.

    Lymph nodes: To "filter" the lymph by removing bacteria, viruses, tumor cells.

30. Lymphatic vessels collect excess tissue fluid (lymph) and return it to the bloodstream. Lymph nodes remove foreign material such as bacteria and tumor cells from lymph and produce lymphocytes.

31. Lymph nodes are most dense in the cervical, axillary, and inguinal areas.

32. The tonsils are particularly important in preventing bacteria and other foreign pathogens from entering the body via the digestive tract. The spleen functions primarily as a RBC "graveyard", i.e., it destroys worn-out RBCs. It also produces lymphocytes and acts as a blood reservoir.

33. Veins that have become twisted and dilated because of incompetent valves. Inactivity (lack of skeletal milking activity against the veins), which allows the blood to pool in the lower extremities; increased pressure that restricts venous return (as in pregnancy and obesity).

ANSWERS TO AT THE CLINIC QUESTIONS

1. Hypertension: abnormally elevated or high blood pressure (generally described as systolic pressure consistently over 140 mm Hg and diastolic pressure consistently over 90 mm Hg in younger adults).

   Arteriosclerosis: "hardening of the arteries", the result of deposit of fatty-cholesterol substances and calcium salts onto the inner walls of the blood vessels. Arteriosclerosis can be a direct cause of hypertension because it decreases the elasticity of the arteries (thereby increasing peripheral resistance).

   Hypertension is often called the "silent killer" because it progresses initially (and often over a prolonged period) without obvious symptoms. Three lifestyle habits that might help prevent cardiovascular disease are regular exercise, a diet low in saturated fats and salt, and a decrease in stress. (Quitting smoking would also help.)

2. She has pulmonary edema. The right side of the heart is still sending blood to the lungs, but the left side of the heart, the systemic pump, is not pumping blood entering its chamber (from the pulmonary circuit) to the systemic circulation. As the pressure increases in the pulmonary vessels, they become leaky, and fluid enters the tissue spaces of the lungs.

3. Incompetence (not stenosis) of the pulmonary semilunar valve. Incompetent valves produce swishing sounds, and the pulmonary semilunar valve is heard at the superior left corner of the heart, as indicated in this question.

4. The compensatory mechanisms of Mrs. Johnson induce an increase in heart rate and an intense vasoconstriction, which allows blood in various blood reservoirs to be rapidly added to the major circulatory channels.

5. (a) Removal of the axillary lymph nodes results in severe localized edema since the lymphatic vessels are also lost. (b) She can expect chronic edema along the arm, although some lymphatic drainage is eventually reestablished by regrowth of the vessels from veins.

6. The left atrium and the posterior portion of the left ventricle.

CLASSROOM DEMONSTRATIONS AND STUDENT ACTIVITIES

Classroom Demonstrations

1. Film(s) or other AV materials of choice.

2. Use a dissectible heart model or a plasticized human heart to show heart structure.

3. Use a dissectible human torso model to point out the major arteries and veins of the body.

4. Obtain and play a recording of normal and abnormal heart sounds to accompany your presentation of valve function and malfunction. ("Interpreting Heart Sounds" is available on free loan from local chapters of the American Heart Association.)

5. If the required equipment is available, demonstrate the recording of an ECG.

6. Bring in a hand-held pinball machine to demonstrate the one-way nature of valves (heart or venous).

Student Activities

1. Demonstrate how apical and radial pulses are taken, and have students practice on each other.

2. Have students practice finding the important pressure points (see Figure 11.13 on p. 328 of the text).

3. Demonstrate the auscultatory method of determining arterial blood pressure, and provide the necessary equipment (sphygmomanometers and stethoscopes) so that students can practice on each other. Allow students to form hypotheses about the effects of body position and exercise on arterial blood pressure, and design and carry out brief experiments to test them.

4. As a point of interest, when studying heart anatomy have students locate their point of maximal intensity (PMI). To locate their PMI, have students press their fingers between the fifth and sixth ribs just below the left nipple, where they can easily feel the beating of the heart where its apex contracts the chest wall.

5. Describe the signs of an impending heart attack and initiate a class discussion on what lifestyle changes might help avert cardiovascular problems.

6. Use a model of the human vascular system or project an unlabeled transparency of the vascular system. Instruct students to call out the names of the vessels as they are indicated.

7. Ask students to think about answers to the following questions for a class discussion:

   a. What factors or events retard venous return?

   b. Why do water and dissolved solutes leave the bloodstream at the arteriole end of a capillary bend and enter the bloodstream at the venule end of the bed?

   c. Why is the elasticity of the large arteries so important? Or, why is arteriosclerosis such a threat?

   d. Why does a body region develop edema after the lymphatics are removed in that area?

8. Offer extra credit for documentation of CPR training during the semester.

AUDIOVISUAL AIDS

Videotapes

1. *Circulation* (CFI, videotape or filmstrip, 24 min., 1988)
2. *Circulation of the Blood* (IM, 24 min., 1991)
3. *Coronary Heart Disease* (PYR, 37 min., 1990)
4. *The Heart* (NET, 30 min., 1990)
5. *The Heart and Blood Vessels: A Review of Normal Morphology and Introduction to Pathology* (HSC, 9 min., 1986)
6. *Heart Dissection and Anatomy* (IM, 14 min., 1989)
7. *The Living Body: Two Hearts That Beat as One* (FHS, VHS or videodisc, 1990)

Computer Software

1. *Human Circulatory System* (EME, disc for MAC/IBM/Apple, 1993)
2. *The Total Heart* (MC, CD ROM for WIN, 1993)

Videodisc

1. *Muscles: Their Structure and Function, Work of the Heart* (EBF)
2. *Respiration, Circulation, and Digestion* (NGS)

12   Body Defenses

LECTURE HINTS

1. Although inflammation is sometimes painful, students need to understand that it is a beneficial, essential defense mechanism that paves the way for proper tissue repair.

    Discuss the symptoms of inflammation and their physiologic cause.

2. Note that complement proteins represent a nonspecific defense, even though they may be activated as part of a specific response to antigen. Explore the effects of complement deficiency.

3. In the humoral response to antibody, plasma cells secrete antibodies that travel in body fluids to reach the site of injury. In the cell-mediated response, on the other hand, T cells migrate to the injured area. An analogy might be made to sending weapons (in the case of humoral immunity) as opposed to sending troops (in the case of cell-mediated immunity).

4. Although immunity represents specific responses to antibody, the system isn't perfect. Cross-reacting antibody forms the basis for some vaccines and is thought to be a contributor to some autoimmune diseases.

5. Point out that immunological memory may or may not be lifelong.

6. Although there are five antibody classes, antibodies with the same antigen specificity can be found in more than one class. To demonstrate this, use board cutouts. Make constant regions in five different colors representing the five antibody classes. Then make a larger number of variable regions that bind to specific antigens (e.g., a purple variable region with a round binding site binding to a round antigen). Combine the constant regions and variable regions in ways that will support your discussion.

7. There are many clinical applications associated with immunity that students will want to discuss, including AIDS, autoimmunity, hypersensitivity, and transplantation.

ANSWERS TO TEXTBOOK REVIEW QUESTIONS
(Questions appear on pp. 380–381)

1. Mucus provides a mechanical barrier for pathogens by trapping them. Mucosae are found on the outer surface of the eye, in the linings of all body cavities open to the exterior, that is, the digestive, respiratory, urinary, and reproductive tracts.

   Lysozyme, an enzyme that destroys bacteria, is found in saliva and lacrimal fluid.

   Keratin, a protein in epithelial membranes, presents a physical barrier to microorganisms on the skin, as well as being resistant to most weak acids and bases and to bacterial enzymes and toxins.

   Acid pH of skin secretions inhibits bacterial growth. Vaginal secretions are also very acidic. Hydrochloric acid is secreted by the stomach mucosa and acts to kill pathogens.

   Ciliated mucosa of the upper respiratory tract sweep dust- and bacteria-laden mucus superiorly toward the mouth, restraining it from entering the lower respiratory passages.

2. Complement is a group of 20 blood proteins that, when activated, causes lysis of the cell to which the antibodies are attached. Other roles of complement include opsonization, inflammatory actions such as stimulating mast cells and basophils to release histamine (which increases vascular permeability), and attracting neutrophils and other inflammatory cells to the area.

3. Interferons are secreted by virus-infected cells. They diffuse to nearby cells and bind to their membrane receptors, interfering with the ability of viruses to multiply within these cells.

4. The immune response is the response of the body to foreign substances.

5. Antigen is a foreign substance (nonself) that is capable of activating an immune response and of interacting with the products (cells and antibodies of that response). An incomplete antigen is a small molecule that connects with our own proteins, unlike a complete antigen, which does not.

6. Humoral immunity is that portion of the immune system that reflects the work of antibodies produced by B cells or their progeny plasma cells. Cell-mediated immunity reflects a cell-mounted attack against antigens that is mediated by T cells.

7. T cells, notably the helper T cells, regulate not only cell-mediated immunity but humoral immunity as well by interacting directly (or indirectly) with B cells.

8. Immunocompetence is the capability of mounting an immune response against a particular antigen. The appearance of antigen-specific receptors on the surface of a B cell or T cell signals that it has developed immunocompetence. T cells are programmed in the thymus; B cells are programmed in the bursa-equivalent organ, probably red bone marrow in humans.

9. Clonal selection involves the rapid cell division of the antigen-activated lymphocyte to produce a "family" or clone of identical cells, all bearing the same antigen-specific receptors. The macrophage is important in clonal selection because it "presents" the antigens to the immunocompetent cells.

10. Members of a B cell clone include B memory cells and plasma cells, which are antibody-producing "factories."

11. Helper T cells are the major regulatory cells; they activate the cytotoxic T cells and are the population disabled by HIV. Killer T cells interact directly with virus-infected (or foreign) cells to kill (lyse) them. Suppressor T cells cause the immune response to wind down when the enemy (antigen) has been destroyed or inactivated.

12. The primary immune response occurs on the first meeting with the antigen; the secondary response is the second and subsequent meeting with the same antigen. The secondary response is faster because all the preparations (clonal selection and memory cells) have been made and immunological memory has been established.

13. An antibody is basically Y-shaped. It consists of two heavy chains and two light chains, each with a constant region and a variable region. The constant regions determine antibody class and where and how the antibody acts in the body. The variable regions form the antigen-binding sites.

14. The five classes of immunoglobulins are IgM, IgG, IgE, IgA, and IgD. IgD is attached to B cell membranes; IgG is most abundant in plasma and crosses placental barriers; IgE is involved in allergic responses; IgM is the first antibody released in the primary response and some are bound to B cell membranes. IgG and IgM bind complement. IgA is present in body secretions that bathe membranes, that is, in tears, saliva, and mucus.

15. Antibodies defend the body by fixing complement and by neutralizing, agglutinating, and precipitating antigens.

16. An allergy is an abnormally vigorous immune response to an otherwise harmless antigen. Acute allergic responses occur and resolve within minutes and are mediated by IgE antibodies bound to mast cells. The mast cells release histamine upon antigen-antibody binding, and histamine causes blood vessels to become leaky. Typical symptoms are runny nose (or asthma), watery eyes, and hives. Delayed-reaction allergies are mediated by T cells and occur hours to days after exposure to the allergen.

17. An autoimmune disease may result from a change in self-antigens, the appearance of self-antigens not previously recognized by the immune system, or the cross-reaction of antibodies produced against foreign antigens with self-antigens.

ANSWERS TO AT THE CLINIC QUESTIONS

1. The infant will develop artificially acquired active immunity.

2. IgA is found primarily in mucus and other secretions that bathe body surfaces. It plays an important role in preventing pathogens from entering the body. Lack of IgA would result in frequent major/minor infections of the sinuses or respiratory tract infections.

3. The mechanisms for the cardinal signs of acute inflammation involve the entire inflammatory process. The inflammatory process begins as a host of inflammatory chemicals are released into the extracellular fluid. They promote vasodilation, allowing more blood to flow into the area, causing a local hyperemia that accounts for the redness and heat of an inflamed area. The liberated chemicals also increase permeability of local capillaries and large amounts of exudate seep from the bloodstream into the tissue space, causing local edema or swelling. The excessive fluid in the extracellular space presses on adjacent nerve endings, contributing to a sensation of pain. The vaccination triggers an immune response that recognizes proteins on the coat of the virus. Since the virus mutates rapidly, and since mutations alter proteins, the immune response induced in a given year is likely to be ineffective a year later.

CLASSROOM DEMONSTRATIONS AND STUDENT ACTIVITIES

Classroom Demonstrations

1. Film(s) or other AV materials of choice.

2. To launch your discussion of the specificity of antigen-antibody interactions, bring in a simple lock and a bunch of keys. Demonstrate that only one of the keys can open the lock.

3. Invite a speaker from the local or state health department to discuss the biological issues related to AIDS.

Student Activities

1. To open a discussion of the inflammatory process, ask if anyone has a cut or an injury that is in the process of healing. If someone does have a reddened swollen area, have the class members observe it, describe all obvious signs, and then provide the underlying reasons for the signs.

2. Ask students to think about answers to the following questions for a class discussion:

   a. Explain why vaccination provides for a lifelong (or at least a long-term) resistance to a particular disease, while an injection of immune serum or gamma globulin provides only temporary protection.

   b. What is the important difference between natural killer cells and cytotoxic T cells?

   c. Why can the helper T cells be called the "managers of the immune system"?

3. Student assignment for class discussion: Look up articles in the newspaper or popular magazines that concern the social, medical, and economic problems surrounding AIDS.

AUDIOVISUAL AIDS

Videotapes

1. *Accident* (FHS, VHS or videodisc, 28 min., 1990)

2. *AIDS: Questions with Answers* (PLP, 15 min., 1991)

3. *AIDS Research: The Story So Far* (FHS, 57 min., 1992)

4. *Allergy and Immunotherapy* (FHS, 26 min., 1990)

5. *The Body Against Disease* (HRM, 30 min., 1990)

6. *Cell Wars* (FHS, 26 min., 1990)

7. *The Clinical Story of AIDS: An Interview with Dr. Paul Volberding* (CVBS, 28 min., 1990)

8. *Dr. Robert Gallo: The Search for the AIDS Virus* (IM, 28 min., 1988)

9. *The Genetics of Transplantation* (MI, 19 min.)

10. *The Human Immune System: The Fighting Edge* (FHS, 52 min., 1988)

11. *The Immune Response* (IM, 22 min., 1989)

12. *Introduction to Immunology—An Overview* (ALT, 30 min., 1981)

13. *It Won't Happen to Me* (CM, 1992)

14. *The Living Body: Internal Defenses* (FHS, VHS or videodisc, 22 min., 1990)

15. *Living with AIDS: An Overview* (FHS, 22 min., 1995)

16. *Living with HIV: New Treatments, New Hopes* (FHS, 22 min., 1995)

17. *The Microbiology of AIDS* (FHS, 10 min., 1992)

18. *The New Immunology* (ALT, 30 min., 1989)

19. *Stress and Immune Function* (FHS, 26 min., 1990)

20. *Transplants: The Immune System at Risk* (PLP, 29 min., 1990)

Videodisc

1. *Immune System and Reproduction* (NGS)

13 The Respiratory System

SUGGESTED LECTURE OUTLINE

LECTURE HINTS

1. Explain to students that the respiratory system includes the organs that are responsible for delivering oxygen to the blood and eliminating carbon dioxide from the blood. Cellular respiration, on the other hand, involves processes in which cells use oxygen and generate carbon dioxide for the production of ATP.

2. Compare the vocal cords to guitar strings when attempting to explain the pitch and intensity of sound.

3. Although the diaphragm and the external intercostal muscles are important in ventilation, it is the diaphragm that accomplishes most of the change in volume of the thorax. Singers learn to control their diaphragm. In days when women wore corsets, their diaphragms—and their breathing—were severely restricted!

4. Students usually believe that their anterior chest wall moves forward because of incoming air and are surprised to learn that it is really expansion of the thorax that allows air to move in.

5. When discussing the importance of surfactant in aiding lung inflation, ask students to blow up balloons that are dry inside and balloons that are wet inside. The surface tension of the water makes it difficult to inflate the balloon.

6. Emphasize that the respiratory center is much more sensitive to changes in pH and carbon dioxide than it is to changes in oxygen. Thus, if the pH falls or if the carbon dioxide concentration increases, hypoventilation will occur. Oxygen levels must fall dramatically in order for hyperventilation to occur. Hemoglobin carries far more oxygen than tissues normally consume, and this constitutes an oxygen reserve that tissues can use if oxygen levels fall slightly.

7. Take some time to discuss the risk factors, symptoms, and consequences of emphysema and lung cancer.

ANSWERS TO TEXTBOOK REVIEW QUESTIONS
(Questions appear on p. 406)

1. To oxygenate the blood while removing carbon dioxide from it.

2. External respiration is the process of gas exchange that occurs between the blood of the pulmonary capillaries and the external environment (alveolar air). Internal respiration is the process of gas exchange that occurs between the blood of the systemic capillaries and the tissue cells.

3. External nares to nasal cavities to nasopharynx to oropharynx to laryngopharynx to larynx to trachea to right or left primary bronchus to secondary (tertiary, etc.) bronchi to bronchioles (to alveolar ducts) to alveolus.

4. The cartilaginous reinforcements keep the trachea patent during the pressure changes that occur during breathing. The incomplete rings of the posterior tracheal surface make it flexible, allowing a food bolus traveling through the posterior esophagus to bulge anteriorly.

5. Primarily in the nasal cavities, but the warming and moistening process continues for the entire length of the respiratory passageways.

6. Mucus serves to trap dust, bacteria, and other foreign debris that manage to enter the respiratory passageways. Cilia are small projections that extend from the luminal cell surface; in the respiratory system, they beat continuously to propel mucus superiorly toward the throat, where it can be swallowed or expectorated. In this way, the debris-laden mucus is prevented from pooling in the alveoli.

7. The right primary bronchus. Because it is straighter and larger in diameter.

8. If either the middle ear or the sinuses are infected, the exudate will drain into the nasal passages and possibly lead to congestion, or "postnasal drip." Conversely, a nasopharyngeal infection can easily spread to the middle ear cavity or the sinuses because of the continuity of their mucosae, thus causing otitis media or sinusitis, respectively.

9. The elastic tissue in the lungs causes the lungs to recoil passively when inspiratory muscles relax. Thus, whereas inspiration is an active process during which the diaphragm and external intercostal muscles must be activated, expiration requires little or no energy expenditure (unless an individual has chronic obstructive pulmonary disease). The pas-

sageways of the lungs are simply conduits to bring air to the alveoli, where functional gas exchange can occur.

10. Their walls are extremely thin (one layer of squamous epithelium plus a basement membrane), and they present an extremely large surface area.

11. TV: Tidal volume; the amount of air inspired or expired during a normal breath.

    ERV: Expiratory reserve volume; the amount of air that can be forcibly exhaled beyond a normal tidal expiration.

    VC: Vital capacity; total exchangeable air. Vital capacity is the largest of these volumes because VC = TV + IRV + ERV.

12. All nonrespiratory air movements are succinctly described on p. 388 of the text.

13. When the diaphragm contracts, it moves inferiorly, thereby increasing the intrathoracic volume in the superior-inferior dimension. The contraction of the external intercostal muscles elevates the rib cage, increasing the intrathoracic volume in the anterior-posterior and lateral dimensions. As the intrathoracic volume is increased, the intrapulmonary pressure decreases.

14. The relaxation of the inspiratory muscles and elastic recoil of the lungs during expiration decreases the intrapulmonary volume and increases the intrapulmonary pressure. This causes gas (air) in the lungs to flow out of the lungs until the gas pressure inside the lungs is equalized with atmospheric pressure.

15. Oxygen is transported bound to hemoglobin within RBCs. Most carbon dioxide is transported in the plasma in the form of bicarbonate ion.

16. Gases diffuse according to their concentration gradients, that is, from an area of their higher concentration to an area of their lower concentration. Venous blood is high in carbon dioxide and low in oxygen compared to alveolar air; thus, carbon dioxide tends to leave the pulmonary blood to enter the alveolar air, and oxygen tends to move from the alveoli into the pulmonary capillary blood. Arterial blood is high in oxygen and low in carbon dioxide; thus, the diffusion gradient in the tissues is opposite to that in the lungs.

17. Medulla (inspiratory and expiratory centers) and pons (apneustic and pneumotaxic centers).

18. Talking, coughing (and other types of nonrespiratory air movements), exercise, and increased body temperature.

19. Decreases in oxygen content of the blood and changes in carbon dioxide blood concentration (leading to increased or decreased blood pH). The latter factor is much more important in respiratory control.

20. Hyperventilation is rapid, deep breathing. During hyperventilation more carbon dioxide is expelled. Since this decreases the carbonic acid content of the blood, the blood pH increases (becomes more alkaline).

21. In emphysema, the individual has problems exhaling due to loss of elasticity of the lungs. Consequently, expiration becomes an active process, and the person is always tired. A barrel chest develops from air retention, but cyanosis is a late sign. In chronic bronchitis, inspiration is a problem because the respiratory passages are narrowed by the inflamed mucosa and excessive mucus. Infections are common because mucus pools in the lungs. Cyanosis occurs early in the disease.

ANSWERS TO AT THE CLINIC QUESTIONS

1. The boy swallowed the safety pin and it entered the respiratory tract. It could probably be found in the right primary bronchus.

2. Voluntary control of breathing is limited by the body's need to obtain oxygen and get rid of carbon dioxide. When these processes are impaired, involuntary controls take over.

3. Alvin has the symptoms of emphysema.

4. The larynx functions to provide an open airway to the trachea and lungs. Edematous swelling of the mucosa of the larynx would close the airway, blocking all air entering the trachea. Tracheotomy is a surgical incision into the trachea through the anterior neck. It allows air to reach the lungs when the larynx is blocked.

5. While swallowing, respiratory passages are blocked off so that food will be forced to enter the digestive tract. The soft palate rises to block the nasopharynx and the epiglottis folds over the opening into the larynx. If swallowing is uncoordinated, these openings could be blocked inappropriately and air flow prevented.

CLASSROOM DEMONSTRATIONS AND STUDENT ACTIVITIES

Classroom Demonstrations

1. Film(s) or other AV materials of choice.

2. Use a human torso model or a model of the respiratory system to point out the organs of this system.

3. Demonstrate the cohesive effect that exists between the two pleural layers (parietal pleura lining the thorax wall and visceral pleura covering the lungs) by putting a drop of water on a glass slide and then covering it with a second glass slide. Show that the slides easily move from side to side but are very difficult to pull apart directly.

4. Use a "model lung" (a bell jar with balloon "lungs" and a rubber "diaphragm") to illustrate the fact that volume changes in the thorax lead to the flow of gases during breathing.

5. Demonstrate the location of the sinuses within the skull.

6. Obtain a fresh lamb or calf lung pluck (lungs plus attached trachea and heart) from a slaughterhouse. Insert a rubber hose snugly into the trachea, and attach the hose to a source of compressed air. Alternately inflate the lungs with the air and then allow them to deflate passively, to illustrate the huge air capacity and elasticity of the lungs.

7. Obtain some animal blood from a veterinary science facility and, via rubber tubing, bubble some air through the blood to demonstrate the color change that occurs when hemoglobin is well oxygenated.

8. Show prepared slides or diagrams of healthy lung tissue and lung tissue from patients with emphysema or lung cancer.

Student Activities

1. Provide tape measures so students can measure the circumference of the rib cage before and after an inspiration.

2. Provide stethoscopes so students can listen to respiratory (breathing) sounds over various regions of their classmate's thorax. For example, bronchial sounds are produced by air rushing through the large passages (trachea and bronchi), whereas the more muffled vesicular breathing sounds are heard over the smallest of the air passageways and the alveoli.

3. If simple hand-held spirometers are available, have students measure their respiratory volumes, particularly tidal volume and vital capacity. Compare the vital capacity of student volunteers of different ages, athletes and nonathletes, smokers and nonsmokers, and men and women.

4. Provide straws, beakers of water, and pH paper. Have students use the straws to blow into the water in the beakers. Since exhaled air contains substantially more carbon dioxide gas than inhaled air, they will be adding carbon dioxide to the water, which will cause it to become acidic. Have them measure the pH of the water at intervals to follow the pH change.

5. Student assignment for class discussion: Be prepared to answer the following questions at the next class session:

   a. Why would it be desirable (and smart) to practice at high altitudes for a few weeks before an athletic event that was to be held in the high mesa of Mexico City?

   b. Why are only slightly higher atmospheric levels of carbon monoxide gas dangerous?

AUDIOVISUAL AIDS

Videotapes

1. *Asthma* (FHS, 19 min., 1990)
2. *Breathing* (FHS, 20 min., 1995)
3. *Breath of Life* (FHS, 26 min., 1989)
4. *Breathing and Talking* (EBF, 8 min., 1987)
5. *Blood Gases: Interpretation* (MI, 28 min., 1986)
6. *Choking: To Save a Life* (EBF, 12 min., 1981)
7. *CPR for Bystanders* (PYR, 27 min., 1986)
8. *How to Save a Drowning Victim: The Heimlich Maneuver* (AM, 17 min., 1985)
9. *The Living Body: Breath of Life* (FHS, 26 min., VHS/BETA, 1990)
10. *New Breath of Life Revisited* (PYR, 24 min., 1981)
11. *The Respiratory System* (EBF, VHS or videodisc, 26 min., 1989)
12. *The Respiratory System: A Review of Normal Morphology and Introduction to Pathology* (HSC, 5 min., 1986)
13. *Respiration* (IM, 29 min., 1990)

Videodiscs

1. *Respiration, Circulation, and Digestion* (NGS)
2. *Respiratory System, Endocrine System* (EBF)

14 The Digestive System and Body Metabolism

SUGGESTED LECTURE OUTLINE

I. ANATOMY OF THE DIGESTIVE SYSTEM (pp. 408–409)
   A. Organs of the Alimentary Canal (p. 408)
      1. Mouth
      2. Pharynx
      3. Esophagus
      4. Stomach
      5. Small Intestine
      6. Large Intestine
   B. Accessory Digestive Organs (pp. 416–418)
      1. Pancreas
      2. Liver and Gallbladder
      3. Salivary Glands
      4. Teeth

II. FUNCTIONS OF THE DIGESTIVE SYSTEM (pp. 419–428)
   A. Overview of Gastrointestinal Processes and Controls (pp. 419–421)
   B. Activities Occurring in the Mouth, Pharynx, and Esophagus (pp. 421–423)
      1. Food Ingestion and Breakdown
      2. Food Movement—Swallowing and Peristalsis
   C. Activities of the Stomach (pp. 423–426)
      1. Food Breakdown
      2. Food Propulsion
   D. Activities of the Small Intestine (pp. 426–427)
      1. Food Breakdown and Absorption
      2. Food Propulsion
   E. Activities of the Large Intestine (pp. 427–428)
      1. Food Breakdown and Absorption
      2. Propulsion of the Residue and Defecation

III. NUTRITION AND METABOLISM (pp. 428–442)
   A. Nutrition (pp. 428–430)
      1. Dietary Sources of Major Nutrients
         a. Carbohydrates
         b. Lipids
         c. Proteins
         d. Vitamins
         e. Minerals
   B. Metabolism (p. 430)
   C. Carbohydrate, Fat, and Protein Metabolism in Body Cells (pp. 430–434)
      1. Carbohydrate Metabolism
      2. Fat Metabolism
      3. Protein Metabolism

D. The Central Role of the Liver in Metabolism (pp. 434–436)
   1. General Metabolic Function
   2. Cholesterol Metabolism and Transport
E. Body Energy Balance (pp. 436–442)
   1. Regulation of Food Intake
   2. Metabolic Rate and Body Heat Production
      a. Basal Metabolic Rate
      b. Total Metabolic Rate
   3. Body Temperature Regulation
      a. Heat-Promoting Mechanisms
      b. Heat Loss Mechanisms

IV. DEVELOPMENTAL ASPECTS OF THE DIGESTIVE SYSTEM AND METABOLISM (pp. 442–443)

LECTURE HINTS

1. The secretions made in various parts of the alimentary canal, the salivary glands, the liver, and the pancreas are derived from blood plasma. Throughout the course of the classes on digestion, keep a running tab of the volume of blood plasma used to generate fluids for digestion. This will help students appreciate the role of the large intestine in reabsorbing water, and help them understand the consequences of diarrhea.

2. Emphasize the structural modifications in each area of the GI tract that lead to optimum digestion and absorption.

3. Use a video to demonstrate peristalsis.

4. Rennin will turn casein (milk protein) into a curd (solid) in an infant's stomach. The liquid that is left is called whey. Adding some rennin, available from biological supply companies, to milk will quickly demonstrate this point. Ask students why they think this might be beneficial in infants.

5. Note that because of vitamin B production, patients with gastrectomies may develop pernicious anemia.

6. When discussing salivary amylase, ask students how bread tastes when it is chewed for a long period of time before it is swallowed. It should taste sweet.

7. The digestive function of the liver is to produce bile. At this time, remind students of the liver's other functions, perhaps having them brainstorm to develop a list. Then introduce new functions related to metabolism.

8. Students will be interested in dental hygiene, dental caries, and changes in the teeth and gums that accompany aging.

9. If time permits, you might want to introduce other topics, such as cirrhosis, peptic ulcers, duodenal ulcers, colon cancer, pancreatic cancer, hepatitis, gallstones, or eating disorders.

ANSWERS TO TEXTBOOK REVIEW QUESTIONS
(Questions appear on pp. 447–448)

1. See Figure 14.1 (p. 409).

2. See Figure 14.1 (p. 409). The arrows should point as follows: from salivary glands into the oral cavity; from the liver and pancreas into the small intestine.

3. Mucosa, submucosa, muscularis, externa, serosa.

4. Mesentery: The peritoneal extension that suspends the alimentary tube organs in the abdominal cavity and provides for entry and exit of nerves and blood vessels to those organs.

   Peritoneum: The double layer of serous membrane that lines the abdominal cavity walls (parietal peritoneum) and covers the exterior of the abdominal cavity organs (visceral peritoneum).

5. Small intestine: Duodenum, jejunum, ileum.

   Large intestine: Cecum, ascending colon, transverse colon, descending colon, sigmoid colon, rectum, anal canal.

6. Villi are fingerlike extensions of the mucosal layer of the small intestine. Within a villus is a capillary bed and a lacteal. Villi increase the surface area of the small intestine for enhanced absorption of foodstuffs.

7. The normal number of permanent teeth is 32; there are 20 deciduous teeth. Enamel covers the tooth crown; dentin makes up the bulk of the tooth.

   Pulp: Connective tissue invested with blood vessels and nerve endings; located within the central pulp cavity of a tooth.

8. Parotid, submandibular, and sublingual. The mucus content of saliva serves to moisten the food and facilitates its compaction into a bolus. The serous component contains the enzyme salivary amylase, which begins the digestion of starch.

9. The bread would begin to taste sweet as the starch is digested to its glucose building blocks.

10. Mouth: The teeth break or tear the food into smaller fragments.

    Stomach: The third obliquely oriented muscle layer in the muscularis externa allows the stomach to physically churn or pummel the contained foodstuffs.

11. Stomach.

12. The protein-digesting enzymes of the stomach (mainly pepsin) are activated and function best at a low pH. The mucus secreted by the stomach glands protects the stomach from self-digestion.

13. Pancreas.

14. Pancreatic juice is the major source of lipases, or fat-digesting enzymes. Bile emulsifies fat, thereby increasing the ability of lipases to digest fatty foods; bile is also necessary for absorption of the products of fat breakdown. Thus, without bile and/or pancreatic juice, little fat breakdown or absorption can occur; the result is fatty stools.

15. Emulsify: To physically break apart larger particles into smaller ones; to spread thin.

16. Gastrin, a hormone produced by the stomach cells, stimulates the stomach glands to produce increased amounts of enzymes, mucus, and hydrochloric acid.

    Secretin, a hormonal product of the small intestine, causes the liver to increase its output of bile and the pancreas to increase its output and release of pancreatic juice.

17. The buccal phase, which takes place in the mouth and is voluntary, consists of the chewing of food and the forcing of the food bolus into the pharynx by the tongue. The involuntary phase, which follows the buccal phase, involves the closure of all nondigestive pathways to the entry of food and the conducting of the food to the stomach by peristaltic waves of the pharyngeal and esophageal walls.

18. Segmental movements are local, rhythmic constrictions of the intestine, which primarily serve to mix the food with digestive juices. Peristaltic movements involve alternate waves of contraction and relaxation of the intestinal walls, by which the food is propelled along the tract.

19. The end-products of protein digestion are amino acids; of fat digestion, glycerol and fatty acids; of carbohydrates, simple sugars (monosaccharides).

20. In the small intestine.

21. Water, some vitamins (K and B), and some ions.

22. Feces, the final product delivered to the rectum, is primarily indigestible food residue and bacteria.

23. Defecation reflex: A cord-mediated reflex that causes the walls of the sigmoid colon and rectum to contract and the anal sphincters to relax when feces enters the rectum.

    Constipation: A situation in which the stool is hard (usually from excessive dehydration) and difficult to pass.

    Diarrhea: A passage of watery stool; generally results from an irritation of the bowel that causes the contents to be propelled along too rapidly for adequate water reabsorption.

24. Metabolism: The sum total of all chemical reactions that occur in the body.

    Anabolism: A metabolic process in which more complex structures or molecules are constructed from simpler ones.

    Catabolism: A metabolic process in which more complex substances are broken down into simpler substances.

25. Gluconeogenesis: The formation of glucose from noncarbohydrate sources, that is, from fats or proteins.

    Glycogenolysis: The breakdown of glycogen to its glucose building blocks.

    Glycogenesis: The formation of glycogen from glucose.

26. Carbohydrates, principally glucose, are most important as fuel sources for ATP production. Proteins are the most important food group for the building of cell structures.

27. When excess fats are oxidized, acidosis is likely. Starvation and diabetes mellitus.

28. BMR: Basal metabolic rate, that is, the amount of energy consumed by the body when it is at rest (basal conditions). Age, sex, body size (and surface area), and the amount of thyroxine produced by the thyroid gland all influence an individual's BMR.

29. You will gain weight.

30. Oxidation of 1 g of carbohydrate produces 4 kcal; 1 g of protein produces 4 kcal; 1 g of fat yields 9 kcal. You would have consumed 290 kcal (80 kcal from protein, 120 kcal from carbohydrate, and 90 kcal from fat).

31. The balance is lost as heat (some of which warms the body tissues and blood).

32. The hypothalamus.

33. Heat is lost from the body by radiation from the skin surface or by evaporation of perspiration from the skin surface. Body heat is retained by withdrawal of blood from the skin capillaries (thus preventing radiation); body heat is generated at an increased rate by the initiation of shivering.

34. Fever: controlled hyperthermia or a body temperature that is controlled at higher-than-usual levels. Fever indicates some type of malfunction, trauma, or infection in the body.

35. Middle-aged adults: Ulcers, gallstones, inflammatory disease of the gallbladder.

    Adolescents: Appendicitis.

    Elderly individuals: Peridontal disease, malabsorption, stomach/colon cancer.

ANSWERS TO AT THE CLINIC QUESTIONS

1. John was suffering from heat exhaustion due to an excessive loss of body fluids (indicated by his wringing wet T-shirt) and his low blood pressure and cool clammy skin. To help his recovery, he should be given fluid and electrolyte therapy and be cooled down.

2. Harry's symptoms indicate a fever caused by his bacterial pneumonia. The white blood cells battling the pneumonia release pyrogens that act directly on the hypothalamus, causing its neurons to release prostaglandins. The prostaglandins reset the hypothalamic thermostat to a higher temperature, causing the body to initiate heat-promoting mechanisms. Vasoconstriction reduces heat loss from the body surface, promotes cooling of the skin, and shivering.

3. Histamine is one of the chemical stimuli for HCl secretion; thus, an antihistamine drug will inhibit HCl secretion; perforation, peritonitis, and massive hemorrhage. She was told not to take aspirin because it can cause stomach bleeding.

4. Leakage of HCl and pepsin from a perforating gastric ulcer will literally erode and digest away other tissues with which these chemicals come into contact.

5. Rickets, a childhood disease in which bones lack calcium salts. Weight-bearing bones, such as the bones of the legs and pelvis, may bend, deform, or break. Milk is a source of vitamin D, that acts as a cofactor to enhance calcium absorption in the small intestine.

CLASSROOM DEMONSTRATIONS AND STUDENT ACTIVITIES

Classroom Demonstrations

1. Film(s) or other AV materials of choice.

2. Use a dissectible human torso model to point out the organs of the digestive system.

3. Use a human skull or a jaw model to demonstrate the different tooth shapes, types, and numbers.

4. Demonstrate the emulsifying action of bile. First mix oil and water together and allow the layers to separate. Then add bile salts and shake vigorously. Point out that the layer of oil has been dispersed into hundreds of tiny fatty spheres by the action of the bile salts.

5. Borrow a specimen jar containing gallstones from a local GI surgeon to show to the class and to enhance your presentation of why gallstones form and why they are very painful.

6. Demonstrate molecular models of carbohydrate, fat, and protein molecules, and show the breaking of their bonds during digestion by enzymes.

Student Activities

1. Have two groups of students obtain information on two different types of diet systems, and a third group investigate fasting as a means of losing weight. Members of each group should come prepared to discuss the benefits and dangers of the diet they have researched.

2. Have students make a list of all the disadvantages of being obese that they can think of.

3. Ask students to think about and answer the following question before the next class: "How do you explain the fact that blood glucose levels remain relatively constant in people that are on a low-carbohydrate diet?"

4. Have students calculate their total caloric intake over a 24-hour period by using a simple caloric guide (obtainable in most drugstores). Then have them analyze their diet with an eye to what improvements could be made to their eating habits.

5. When discussing the swallowing mechanism, have students place their hands on their larynx so that they can feel it rise when they swallow. You might want to provide small cups of water.

AUDIOVISUAL AIDS

Videotapes

1. *Animal Nutrition* (IM, 29 min., 1990)
2. *The Appendix* (PLP, 10 min., 1989)
3. *Cellular Respiration: Energy for Life* (HRM, 30 min., 1994)
4. *Cholesterol Explained* (FHS, 19 min., 1995)
5. *Contemporary Nutrition* (FHS, VHS or videodisc, 60 min., 1995)
6. *Digestion* FHS, 20 min., 1995)
7. *Digestive Disorders* (FHS, 19 min., 1989)
8. *The Liver* (IFB, 15 min., 1979)
9. *The Living Body: Breakdown* (FHS, VHS or videodisc, 26 min., 1989)
10. *The Living Body: Eating to Live* (FHS, VHS or videodisc, 26 min., 1989)
11. *The Living Body: Hot and Cold* (FHS, VHS or videodisc, 26 min., 1989)
12. *Metabolism: Structure and Relation* (EIL, 7 min.)
13. *Metabolism: The Fire of Life* (EIL, 35 min., 1982)
14. *Nutrition: The All-American Meal* (BARR, 11 min., 1988)
15. *Nutrition: Eat and Be Healthy* (MF, 18 min., 1994)
16. *Nutrition for Living* (FHS, VHS or videodisc, 59 min., 1995)
17. *Nutritional Aspects of the AIDS Patient* (ALT, 29 min., 1989)
18. *Regulating Body Temperature* (EBF, 22min., 1972)
19. *The Stomach and Its Disorders* (PLP, 10 min., 1989)

Videodiscs

1. *Regulating Body Temperature, The Digestive System* (EBF)
2. *Respiration, Circulation, and Digestion* (NGS)

15   The Urinary System

LECTURE HINTS

1. Explain to students that the kidneys are not only the primary organs for the elimination of wastes, they also control the plasma concentration of other substances such as water and electrolytes. This control is usually under the influence of hormones, such as parathyroid hormone for calcium and antidiuretic hormone for water.

2. Although students realize that kidneys are excretory organs, they forget about the other functions of the kidneys. Ask them what other problems a dialysis patient might have. Hopefully they will remember that the kidneys regulate the rate of red blood cell production, are important in the activation of vitamin D, and have an impact on blood pressure and pH.

3. Make a chart of normal and abnormal constituents of urine.

4. When discussing filtration and tubular reabsorption an analogy can be made to cleaning closets. One very effective way to clean out a closet is to remove everything (filtration), and then put back into the closet only what you want to keep (reabsorption). The rest is thrown away.

5. A head of lettuce sitting in a bowl, or a baseball in a baseball glove, can be used to represent the relationship between the glomerulus and Bowman's capsule.

6. If time permits, discuss kidney transplantation, dialysis, chronic renal failure, cystitis, glomerulonephritis, or diabetes insipidus.

ANSWERS TO TEXTBOOK REVIEW QUESTIONS
(Questions appear on p. 470)

1. Kidneys: Formation of urine, regulation of acid-base balance, electrolyte balance, and water balance of the blood.

   Ureters: Conduct urine from the kidneys to the bladder.

   Bladder: Temporary storage of urine.

   Urethra: Conduct urine from the bladder to the body exterior.

2. The kidneys are retroperitoneal in the superiodorsal lumbar region of the body.

3. See Figure 15.2 (p. 452).

4. The nephron.

5. Glomerulus to glomerular (Bowman's) capsule to proximal convoluted tubule, through loop of Henle and distal convoluted tubule to the collecting duct (which leaves the cortex of the kidney and traverses the renal columns and medullary region to empty into a calyx, which in turn empties into the kidney pelvis) to ureter to bladder to urethra.

6. The glomerulus is the "filter"; blood fluids and substances smaller than proteins pass out of the blood at the glomerulus into the renal tubules. The renal tubules reabsorb useful substances (such as amino acids, glucose, and certain ions) so that they can be shunted back into the bloodstream. They also actively secrete various substances, such as hydrogen ions, creatinine, and drug metabolites into the filtrate.

7. The kidney controls water balance, ion levels, and the acid-base balance of the blood.

8. Filtrate is equivalent to blood plasma without the plasma proteins. Urine is "processed filtrate," that is, filtrate from which useful substances (amino acids, glucose, needed ions, and water) have been reabsorbed into the blood.

9. Aldosterone causes the kidney tubule cells to reabsorb more sodium ions (and to secrete potassium ions). As the sodium ions are reabsorbed, water follows. Thus, aldosterone causes the removal of sodium ions (and water, secondarily) from the filtrate.

10. Antidiuretic hormone (ADH). Diabetes insipidus, a condition in which huge volumes of water are lost from the body into the urine, resulting in severe dehydration.

11. Glucose, proteins, red blood cells, white blood cells. Conditions arising from the presence of each of these substances in the urine are (in order): glycosuria, proteinuria, hematuria, pyuria.

12. The presence of abnormal substances in the urine is diagnostic of a pathology and may indicate a problem before other symptoms are present.

13. Micturition: Voiding, or emptying of the bladder (so that urine is flushed from the body).

   Micturition reflex: When approximately 200 mL of urine have collected in the bladder, strength receptors in its wall are activated. Impulses, transmitted to the spinal cord and then back to the bladder (a reflex arc), cause the bladder to contract, and urine is forced past the internal sphincter. It is then that one feels the urge to void. If the time is inconvenient, voiding can be prevented by voluntary closure of the external sphincter and, within a few seconds, bladder contractions cease. After 200 to 300 mL more urine have collected in the bladder, the reflex is initiated once again.

14. Prolonged stasis of urine in the bladder can result in calculi and encourages infectious disease of the bladder.

15. Incontinence: Inability to voluntarily control the external urethral sphincter.

16. The female urethra is much shorter (1.5 inches vs. 8 inches in the male). The female urethra conducts urine only, whereas the male urethra conducts both sperm and urine.

17. Because the urethra is short and lies close to the posterior anal body orifice. Improper toilet hygiene, that is, wiping back to front, carries colonic bacteria to the urethra, thereby promoting infection of the normally sterile urinary passages.

18. Infections.

19. The efficiency of the kidneys (particularly in relation to their ability to form concentrated urine) declines as arteriosclerosis plagues the kidney circulation and as the renal tubules begin to deteriorate. A loss of bladder size and tone leads to increasing episodes of urgency, frequency, and incontinence. In males, prostatic hypertrophy may lead to urinary retention accompanied by calculi and frequent cystitis.

ANSWERS TO AT THE CLINIC QUESTIONS

1. Renal calculi (kidney stones). Predisposing conditions are frequent bacterial infections of the urinary tract, urinary retention, and alkaline urine. The woman's pain comes in waves because the stone is probably passing through the ureter. Pain occurs when the ureter walls close in on the sharp calculi as they are being eased through the ureter by peristalsis.

2. Acute glomerulonephritis due to a reaction to streptococcus bacteria.

3. Urinary tract infection (urethritis and cystitis), probably by E. coli.

4. Pituitary; hyposecretion of ADH; diabetes insipidus.

5. Parathyroid hormone (PTH), released from the parathyroid gland, enhances the reabsorption of calcium by kidney tubules, and decreases the retention of phosphates. Hypersecretion of PTH would cause elevated blood calcium (hypercalcemia).

CLASSROOM DEMONSTRATIONS AND STUDENT ACTIVITIES

Classroom Demonstrations

1. Film(s) or other AV materials of choice.

2. Use a dissectible human torso model or a model of the urinary system to point out the organs of this system.

3. Use a model of a longitudinally sectioned kidney to identify its major anatomical regions: cortex, medulla, pelvis, and associated blood supply. If the nephron is shown as a part of the model, also demonstrate the anatomic regions of the nephron and describe their specific functions in filtrate formation or processing.

4. Display a hydrometer and other materials used to perform a urinalysis, and discuss the importance of urinalysis in both routine physicals and pathological diagnoses. Simulated urine is available from several biological supply houses and eliminates the need for students to provide their own samples.

5. Redemonstrate filtration, osmosis, and diffusion as a reminder of the mechanisms of these processes. (See demonstrations 5, 7, and 8 in Chapter 3.)

6. If possible, arrange for someone from a local renal dialysis center to come and talk to the class about how the artificial kidney works and other aspects of the dialysis process.

Student Activities

1. Have students research the thirst mechanism and be prepared to discuss it in class.

2. Request that the students talk with a nurse at a local hospital to find out why I & O (fluid intake and output) records are so carefully obtained in very ill patients. Ask that they be prepared to communicate that information to the class.

3. Assign students to determine the answer to the following question for the next class: "What is the role of the kidneys in blood pressure regulation, and how do the kidneys accomplish this role?"

AUDIOVISUAL AIDS

Videotapes

1. *Acid/Base Balance: The Ups and Downs of pH* (CM, 15 min., 1990)

2. *Body Fluids: The Critical Balance, Part 1* (CM, 17 min., 1990)

3. *Kidney Disease* (FHS, 26 min., 1992)

4. *Kidneys: The Remarkable Filters* (BFA, 13 min., 1989)

5. *The Living Body: Water* (FHS, 26 min., 1989)

6. *The Mammalian Kidney* (EIL, 60 min., 1987)

7. *Water* (FHS, VHS or videodisc, 26 min., 1990)

8. *Work of the Kidneys* (EBF, 20 min., 1972)

16   The Reproductive System

SUGGESTED LECTURE OUTLINE

D. Childbirth (pp. 493–496)
   1. Initiation of Labor
   2. Stages of Labor
      a. Stage 1: Dilation Stage
      b. Stage 2: Expulsion Stage
      c. Stage 3: Placental Stage

VII. DEVELOPMENTAL ASPECTS OF THE REPRODUCTIVE SYSTEM (pp. 496–498)

LECTURE HINTS

1. The female and male reproductive systems consist of gonads, tubes that transport gametes, and glands. Make a chart to compare the male and female reproductive systems in these categories.

2. Discuss the mechanisms that are used to keep the testes cool for spermatogenesis. Indicate that if the testes fail to descend, infertility may result. There is a higher incidence of testicular cancer in these men as well.

3. Have students investigate and report on causes of male or female infertility.

4. Show a video that discusses prenatal growth and development.

5. Trace the pathway that sperm takes from its origin to its union with the egg in the fallopian tube. Indicate where secretions are added and what the functions of the secretions are. Cards listing each structure can be placed on the board in random fashion and unscrambled by the students.

6. If time permits, discuss sexually transmitted diseases in some detail. Other topics, such as contraception, sterilization, in vitro fertilization, prostate cancer, ovarian cancer, cervical cancer, breast cancer, hormone replacement therapy, and the effects of aging on the reproductive system can also be discussed.

ANSWERS TO TEXTBOOK REVIEW QUESTIONS
(Questions appear on p. 502)

1. The testes are the male gonads. They form sperm and testosterone.

2. Testis to epididymis to ductus (vas) deferens to the ejaculatory duct to the urethra, which extends from the prostrate interior to the tip of the penis.

3. To provide a liquid medium for carrying sperm out of the male body and into the female reproductive tract. Prostate gland, seminal vesicles, and bulbourethral glands.

4. Erection.

5. Ejaculation: Propulsion of sperm (in seminal fluid) out of the male's body.

6. Internal body temperature is too high for the production of viable human sperm. The lower temperature that is required is provided for by the testes being housed in the scrotum, a divided skin sac that hangs outside the body cavity, posterior to the penis.

7. Hypertrophy of the prostate gland, which surrounds the urethra, constricts the urethral passageway, thus inhibiting the passage of semen.

8. The seminiferous tubules of the testes form sperm. Spermatogenesis begins during puberty under the influence of FSH from the anterior pituitary. LH (another anterior pituitary hormone) causes the testes to produce testosterone at that time. Testosterone is also necessary for normal sperm production.

9. The extraneous cytoplasm is sloughed off the spermatid and the nucleus, and remaining organelles are compacted into three major regions so that a streamlined sperm with the power of locomotion is produced. This process is called spermiogenesis.

10. Increased hair growth all over the body and particularly on the face, axillary and pubic regions; deepening of the voice; enlargement of the skeletal muscles; thickening of the bones.

11. The ovary. It produces eggs (actually oocytes) and the female sex hormones estrogen and progesterone.

12. Uterine (Fallopian) tubes: The usual site of fertilization; conduct the oocyte (or embryo) to the uterus.

    Uterus: The site of implantation of the embryo.

    Vagina: The birth canal and the chamber through which menstrual flow leaves the body.

13. The distal ends of the uterine tubes have fingerlike extensions called fimbriae, which wave and create fluid currents to draw the ovulated body into the uterine tube.

14. FSH promotes follicle development. LH cooperates with FSH in follicle maturation and triggers ovulation.

    Follicle: An immature egg (oocyte) surrounded by one or more layers of smaller cells called follicle cells.

    Ovulation: Ejection of an oocyte (surrounded by a capsule of follicle cells) from the ovary.

15. The human oocyte becomes an ovum only if sperm penetration occurs, at which time the secondary oocyte completes the second meiotic division to produce the ovum nucleus and a polar body, which is ejected.

16. Estrogen, because it promotes the appearance of the secondary sex characteristics of the female. Both growing and mature (Graafian) ovarian follicles form estrogen. The second hormone produced by the ovary is progesterone.

17. See pp. 484–486 of the text. It provides the uterine environment needed for implantation.

18. Menopause: The period during which a woman's ovarian function (thus, reproductive ability) gradually declines and finally ends.

19. To produce milk to nourish the infant.

20. Fertilization: Fusion of the sperm and egg nuclei to form the nucleus of the fertilized egg, or zygote. It normally occurs in the uterine tube of the female. Implantation involves the erosion of the uterine mucosa by the embryo, and then the covering over of the embedded embryo by mucosa regrowth.

21. The placenta serves to deliver nutrients and oxygen to the embryo/fetus from the mother's circulation and to remove carbon dioxide and other metabolic wastes from the embryo's blood (into the mother's blood); it also forms the hormones of pregnancy (estrogen, progesterone, and others). Thus, in addition to performing endocrine functions, the placenta is a respiratory, excretory, and nutritional organ for the developing fetus.

22. During pregnancy the woman's respiratory and urinary systems must function "for two." Respiration becomes increasingly difficult as the uterus enlarges and presses the abdominal organs superiorly against the diaphragm. Cardiac output and blood pressure also increase to meet the increased circulatory demands. Changes in the woman's weight distribution and center of gravity make walking more difficult and lead to muscle strain. Urinary frequency sometimes occurs as the uterus presses on the bladder.

23. Except for prostatic hypertrophy, aging males seem to have few age-associated reproductive system problems. Reproductive capacity in females ends during menopause. Problems associated with the relative lack of female hormones after menopause include high blood pressure, bone loss, and predisposition to vaginal infections.

ANSWERS TO AT THE CLINIC QUESTIONS

1. The woman is in the second, or expulsion, stage of labor. Since this stage typically takes 50 minutes for the first birth and 20 minutes in subsequent births, she probably would not have time to get to the hospital.

2. It is possible for an oocyte to be ovulated from a right ovary into the peritoneal cavity, and then move through the fluid of the peritoneal cavity all the way to the infundibulum of the left uterine tube. Several cases have been documented of women like Lucy bearing children.

3. Adolf probably has a tumor of the prostate gland, and the enlargement of this gland is constricting his prostatic urethra. In the rectal exam, a physician is feeling the prostate for swelling and lumps.

4. The mucus produced by these glands cleans the urethra of urine residues.

5. Her only options are coitus interruptus, or withdrawal of the penis just before ejaculation, or rhythm or fertility awareness methods, based on recognizing the period of ovulation or fertility and avoiding intercourse during those intervals.

6. Ibuprofen is a prostaglandin inhibitor; such drugs can inhibit labor in the early stages.

CLASSROOM DEMONSTRATIONS AND STUDENT ACTIVITIES

Classroom Demonstrations

1. Film(s) or other AV materials of choice.

2. Use a dissectible human torso model to point out the organs of the reproductive system.

3. If available, use a "pregnant" torso model to illustrate the most desirable positioning of the placenta and the vertex presentation of a fetus.

4. Use models showing the process of meiosis in spermatogenesis and oogenesis.

5. Use embryonic-fetal development models to show the stage of cleavage, the blastula, etc.

6. If available, show preserved specimens of human embryos or fetuses in various stages of development to illustrate the later changes that occur during gestation.

Student Activities

1. Have students find and bring in an article published in the newspaper or a popular magazine that deals with the effects of maternal drug abuse or disease (e.g., AIDS, herpes) on the well-being of the fetus.

2. Student assignment for class discussion:

   a. Define episiotomy and explain why this procedure may be performed.

   b. Define Down's syndrome and indicate in which maternal age group it is most common and why.

   c. Explain why new mothers sometimes experience abdominal cramping when nursing their babies.

3. Display various types of contraceptive techniques as an adjunct to a discussion of birth control.

4. After your demonstration, have students practice examining their breasts for lumps. Alternatively, distribute a handout showing the correct breast examination procedure. Assign this activity as a monthly priority for both males and females.

5. Have students research and report on the following topics:

   a. Antenatal disease and treatment.

   b. In vitro fertilization.

   c. Aging of the reproductive system.

   d. Breast cancer.

   e. Contraceptive methods for males and females.

AUDIOVISUAL AIDS

Videotapes

1. *AIDS: Myths and Facts* (MFM, 22min., 1994)

2. *Breast Cancer: Working Towards a Cure* (FHS, 21 min., 1995)

3. *Coming Together* (FHS, VHS or videodisc, 26 min., 1990)

4. *Heredity, Health, and Genetic Disorders* (HRM, 1982)

5. *Human Embryology Series* (TFI, six 15 min. films, 1990)

6. *Human Reproduction* (CF, 20 min., 1989)

7. *Infertility: New Treatments* (FHS, 25 min., 1995)

8. *Into the World* (FHS, VHS or videodisc, 28 min., 1990)

9. *Life in the Womb: The First Stages of Human Development* (HRM, 40 min., 1989)

10. *The Living Body: Shares in the Future* (FHS, VHS or videodisc, 26 min., 1990)

11. *Meiosis—The Key to Genetic Diversity* (HRM, 26 min., 1991)

12. *The Miracle of Life* (PBS, 60 min., 1986)

13. *A New Life* (FHS, VHS or videodisc, 28 min., 1990)

14. *Reproduction: Designer Babies* (FHS, VHS or videodisc, 20 min., 1995)

15. *The Reproductive System—A Baby is Born* (MFM, 45 min., 1989)

16. *Sexually Transmitted Diseases* (PLP/FHS, 19 min., 1990)

17. *STDs and Safer Sex: Your Risk and Responsibility* (BC, 20 min., 1992)

18. *Vasectomy* (PLP, 10 min., 1989)

19. *A Woman's Body* (FHS, 49 min., 1990)

Videodiscs

1. *Human Sexuality Videodisc* (HRM, 120 min., 1992)

2. *Immune System and Reproduction* (NGS)

Audiovisual Resources Guide

AC   Annenberg/CPB Collection, Intellimation, PO Box 1922, Santa Barbara, CA 93116-1922. (805) 968-2291.

ACR   American College of Radiology, 1891 Preston White Dr., Reston, VA 22091-4397. (703) 648-8900.

ACS   American Cancer Society, 19 W. 56th St., New York, NY 10019. (212) 586-8700.

ADAM   ADAM Software Inc. 1899 Powers Ferry Road, Suite 460, Marietta, GA 30067

AEF   American Educational Films, PO Box 70188, Nashville, TN 37207. (615) 868-2040.

AHP   Alfred Higgins Productions, 6350 Laurel Canyon Blvd., Suite 305, North Hollywood, CA 91606. (818) 762-3300.

AIF   Australian Instructional Films, 39 Pitt St., Sydney, Australia.

AJN   The American Journal of Nursing Co., Educational Services Division Film Library, 555 W. 57th St., New York, NY 10019. (800) 223-2282.

ALT   Altschul Group Corporation, 1560 Sherman Ave., Evanston, IL 60201. (See TFI.)

AM   AIMS Media, 9710 DeSoto Ave. Chatsworth, CA 91311-4409. (818) 773-4300

AMA   American Medical Association, 515 N. State St., Chicago, IL 60610. (800) 621-8335.

ASFT   Association-Sterling, 8615 Directors Row, Dallas, TX 75240.

AVC   Audio Visual Concepts, Isabel Andreu Aguilar, Suite 106, Hato Rey, Puerto Rico 00918. (809) 753-7700.

BARR   Barr Films, 12801 Schabarum, P.O. Box 7878, Irwindale, CA 91706-7878. (818) 338-7878.

BC   Benjamin/Cummings Publishing Company, 390 Bridge Parkway, Redwood City, CA 94065. (800) 447-2226.

BFA   BFA Educational Media, Division of Phoenix Learning Group, 2349 Chaffee Dr., St. Louis, MO 63146. (800) 221-1274.

BL   Bristol-Myers-Squibb, P.O. Box 4755, Syracuse, NY 13221. (315) 432-2000.

BNF   Benchmark Films, Inc., 145 Scarborough Rd., Briarcliff Manor, NY 10510. (914) 762-3838.

BYU   Brigham Young University, Media Services, 101 Fletcher Building, Provo, UT 84602. (801) 378-4071.

CBS   Carolina Biological Supply Company, 2700 York Rd., Burlington, NC 27215. (919) 584-0381.

CF   Churchill Films, Inc., 12210 Nebraska Ave., Los Angeles, CA 90025. (310) 207-6600.

CFI   Center for Humanities, Inc. Communications Park, Box 1000, Mount Kisco, NY 10549. (914) 666-4100.

CG   The Cinema Guild, 1697 Broadway, Suite 506, New York, NY 10019. (212) 246-5522.

CHM   Cleveland Health Education Museum, 8911 Euclid Ave., Cleveland, OH 44106 (216) 231-5010.

CIBA   Ciba Pharmaceutical Company, Medical Communications Dept., 556 Morris Ave., Summit, NJ 07901. (908) 277-5000.

CIF   Coronet/MTI Film & Video, 108 Wilmot Rd., Deerfield, IL 60015. (800) 621-2131.

CM   Concept Media, 2493 DuBridge Ave., Irvine, CA (714) 660-0727.

CRM   CRM Films, 2215 Faraday Ave., Carlsbad, CA 92008. (800) 421-0833.

CVBS   Connecticut Valley Biological Supply Company, Inc. PO Box 326, 82 Valley Rd., Southampton, MA 01073. (800) 628-7748.

EBF   Encyclopedia Britannica Educational Corp., 310 South Michigan Ave., Chicago, IL 60604. (312) 321-7322.

EHSL   Spencer S. Eccles Health Sciences Library, University of Utah, Building #589, Salt Lake City, UT 84112. (801) 581-8772.

EIL   Educational Images Ltd., P.O. Box 3456, Elmira, NY 14905. (607) 732-1090.

ELE   Lilly & Company, Medical Division, Lilly Corporate Center, Indianapolis, IN 46265. (317) 276-2000.

EME   EME Coproration, P.O. Box 2805, Danbury, CT 06813-2805. (800) 848-2050. (203) 798-2050

FAD   F. A. Davis Company, 1915 Arch St., Philadelphia, PA 19103. (215) 568-2270.

FHS   Films for the Humanities and Sciences, Inc., Box 2053, Princeton, NJ 08543 (800) 257-5126.

FSCE   FSC Educational, 941 Hickory Ln., Mansfield, OH 44905. (419) 589-2100.

HP   Hoechst Roussel Pharmaceuticals, Rt. #202-206-N, Somerville, NJ 08876. (908) 231-2000.

HR   HarperCollins Publishers, 10 E. 53rd St., New York, NY 10022. (212) 207-7000.

HRM   Human Relations Media, 175 Tompkins Ave., Pleasantville, NY 10570. (914) 747-0177.

HSC   Health Science Consortium, 201 Silver Ceda Ct., Chapel Hill, NC 27514-1517. (919) 942-8731

HV   Hawkhill Video, 125 East Gilman Street, P.O. Box 1029, Madison, WI 53701-1029. (800) 442-4295.

ICIA   ICI Americas, Inc., Concord Plaza, 3411 Silverside Rd., Wilmington, DE 19899. (302) 886-3000.

IFB   International Film Bureau, 332 S. Michigan Ave., Chicago, IL 60604. (312) 427-4545.

IM   Insight Media, 121 West 85th St., New York, NY 10024. (800) 233-9910.

IOWA   Media Production Unit, Iowa State University, Exhibit Hall South, IA 50011. (515) 294-2316.

IU   Indiana University, Instructional Support Services, Bloomington, IN 47405-5901. (812) 855-2853.

JBL   J. B. Lippincott Company, 227 East Washington Square, Philadelphia, PA 19106. (215) 238-4444.

JJ   Johnson & Johnson, Grandview Rd., Skillman, NJ 08558. (908) 874-1000.

JW   John Wiley & Sons, Inc., 605 Third Ave., New York, NY 10158. (212) 850-6000.

KSU   Kent State University Audiovisual Center, 330 Library, Kent, OH 44242. (800) 338-5718.

MC   Mayo Clinic, Section of Photography, South 9 Plummer, 200 First St., Rochester, MN 55901. (507) 282-2511.

MF   Milner-Fenwick, Inc., 2125 Greenspring Dr., Timonium, MD 21093. (800) 638-8652.

MFM   Marshfilm/Marshmedia, P.O. Box 8082, Shawnee Mission, KS 66208. (816) 523-1059.

MG   Media Guild, 11722 Sorrento Valley Road, Suite E, San Diego, CA 92121.
        (619) 755-9191.

MGH   Massachusetts General Hospital, Television Department, 55 Fruit St., Boston, MA
        02114. (617) 726-2000.

MGHBC   McGraw-Hill Book Company, Continuing Education Program, 4001 Constitution NW,
        Lower Level, Washington, DC 20008. (202) 244-1600.

MGHCS   McGraw-Hill College Software, 1221 Avenue of the Americas, New York, NY 10020.
        (212) 512-6349.

MI   Medcom, Inc., PO Box 3225, Garden Grove, CA 92642. (714) 891-1443.

MTI   MTI Home Video, 14216 SW 136th St., Miami, FL 33186. (305) 255-8684.

NAC   National Audiovisual Center, National Archives and Records Administration,
        Customer Services, 8700 Edgeworth Dr., Capitol Heights, MD 20743. (301) 763-1896.

NDM   New Dimension Media, 85803 Lorane Highway, Eugene, OR 97405. (503) 484-7125.

NET   Nebraska Educational Television, Council for Higher Education, P.O Box 83111,
        Lincoln, NE 68501. (402) 472-3611.

NGS   National Geographic Society, 1145 17th St. NW, Washington, DC 20036.
        (202) 857-7000.

NTA   National Teaching Aids, 1845 Highland Ave., New Hyde Park, NY 11040.
        (516) 326-2555.

NYAM   New York Academy of Medicine, 2 East 103rd St., New York, NY 10029.
        (212) 876-8200.

ODC   Optical Data Corporation, 30 Technology Dr., Warren, NJ 07059. (908) 668-0022.

PAR   Paramount Communications, c/o AIMS Media, 9710 DeSoto Ave., Chatsworth, CA
        91311. (818) 773-4300.

PBS   PBS Video, 1320 Braddock Place, Alexandria, VA 22314. (800) 424-7963.

PHM   Prentice-Hall Media, Guidance Associates, PO Box 1000, Mount Kisco, NY 10549.
        (800) 431-1242.

PLP   Projected Learning Programs, Inc., PO Box 3008, Paradise, CA 95967-3008.
        (800) 248-0757.

PMR   Peter M. Robeck and Company, Inc., Distributed by Time-Life Films, Inc., 20
        Beckman Place, New York, NY 10021. (212) 371-5388.

PYR   Pyramid Film & Video, PO Box 1048, Santa Monica, CA 90406. (213) 828-7577.

RL   Roche Biomedical Lab, Inc., 1447 York Court, Burlington, NC 27215.

SC   Schering-Plough Research Institute, Information Center, 2015 Galloping Hill Rd.,
        Kenilworth, NJ 07033. (201) 558-4000.

SEF   Sterling Educational Films, 241 E. 34th St., New York, NY 10016. (212) 262-9433.

SM   Science and Mankind, Inc., Communications Park, Box 2000, Mount Kisco, NY
        10549. (914) 666-4100.

TC   Trainex Corp., P.O Box 116, Garden Grove, CA 92542. (714) 898-2561. (See MI.)

TFI   Teaching Films, Inc., Altschul Group Corp., 1560 Sherman Ave., Suite 100, Evanston,
        IL 60201. (708) 328-6700.

TLV   Time-Life Video & Television, Division of Time-Life Inc., 777 Duke St., Alexandria,
        VA 22314. (703) 838-7000.

UCEC	University of California Extension, Center for Media and Independent Learning, 2000 Center St., Fourth Floor, Berkeley, CA 94704. (510) 642-0460

UIFC	University of Illinois Film Center, 1325 South Oak St., Champaign, IL 61820. (217) 333-1360.

UM	University Media, 11526 Sorrento Valley Rd., Suite J, San Diego, CA 92121.

UT	University of Texas Medical Branch, Medical Illustrations Department, Galveston, TX 77555-1036. (409) 772-1011.

UWM	University of Washington, Instructional Media Services, DG-10, Kane Hall, Seattle, WA 98195. (206) 543-9909.

VI	Videodiscovery, Inc., 1700 Westlake Ave. N, Suite 600, Seattle, WA 98109-3012. (800) 548-3472, (206) 285-5400.

WARDS	Ward's Natural Science Establishment, Inc., PO Box 92912, Rochester, NY 14692. (716) 359-2502

WAYNE	Wayne State University, Media Services, 5265 Cass Ave., Detroit, MI 48202. (313) 577-1980.

ADDITIONAL AUDIOVISUAL RESOURCES

Boucher, Brian, et al. *Handbook and Catalogue for Instructional Media Selection*. Englewood Cliffs, NJ: Educational Technology Publications, 1973.

Crowley, Leonard V. *Visual Aids in Anatomy and Physiology*. Chicago: Yearbook Medical Publishers, 1976.

Educational Media Council: Educational Media Index. New York: McGraw-Hill, 1964.

Educators Guide to Free Films, 52nd rev. ed. Randolph, MA: Educators Progress Service, 1992.

Ely, Donald P., and Barbara B. Minor, eds. *Educational Media and Technology Yearbook*. Englewood, CO: Libraries Unlimited, Inc., 1992.

Hanson, Patricia K., and Stephen L. Hanson, eds. *Film Review Index*. Phoenix, AZ: The Oryx Press, 1987.

Hart, T. L., et al. *Multi-Media Indexes, Lists and Review Sources: A Bibliographic Guide*. New York: Dekker, 1975.

Index to Instructional Media Catalogues. Ann Arbor, MI: Bowker, 1974.

Multi-Media Reviews Index. Ann Arbor, MI: Preirian Press, 1977.

National Information Center for Educational Media (NICEM). *Media Indexes*. University Park, CA: University of Southern California.

Rufsvold, Margaret. *Guides to Educational Media*. 4th ed. Chicago: American Library Association, 1977.

Sive, Mary R. *The Complete Media Monitor: Guide to Learning Resources*. Metuchen, NJ: Scarecrow, 1981.

Sive, Mary R. *Educators' Guide to Media Lists*. Littleton, MA: Libraries Unlimited, 1975.

Taggart, Dorothy T. *A Guide to Sources in Educational Media and Technology*. Metuchen, NJ: Scarecrow, 1975.

Wittrich, Walter A., and James L. Berget. *Educators' Guide to Free Audio and Video Materials*. 24th rev. ed. Randolph, MA: Educators Progress Service, 1977.

Filmstrips and Film Loops

BFA Educational Media, Division of Phoenix Learning Group, 2349 Chaffee Dr., St. Louis, MO 63146. (800) 221-1274.

Carolina Biological Supply Company, 2700 York Rd., Burlington, NC 27215. (919) 584-0381.

Scientific American, Inc., 415 Madison Ave., New York, NY 10017. (212) 754-0550.

Transparencies

Carolina Biological Supply Company, 2700 York Rd., Burlington, NC 27215. (919) 584-0381.

Denoyer-Geppert Audio-Visuals, 5235 Ravenswood Ave., Chicago, IL 60640. (312) 561-9200.

Hammond, Inc., Educational Division, 515 Valley St., Maplewood, NJ 07040. (201) 763-6000.

Photo & Sound Co., 2625 Yates Ave., City of Commerce, CA 90040. (213) 887-9000.

Standard Publishing Co., 8121 Hamilton Ave., Cincinnati, OH 45231. (513) 931-4050.

Videotapes

Multimedia Entertainment, 45 Rockefeller Plaza, New York, NY 10011. (212) 332-2000.

Polymorph Films, 118 South St., Boston, MA 02111. (617) 542-2004.

Part II Test Bank

1   The Human Body: An Orientation

Answer the following multiple-choice questions by circling the letter of the correct response. For matching questions, write the correct response in the answer blank.

(b)   1. All of the following activities represent anatomical studies EXCEPT:

         (a)   making a section through the heart to observe its interior.

         (b)   drawing blood from laboratory animals to determine their blood sugar levels.

         (c)   examining the surface of a bone.

         (d)   viewing muscle tissue through a microscope.

(d)   2. Physiology is generally studied:

         (a)   in terms of chemical and physical events in living systems.

         (b)   to learn about the function of the body.

         (c)   in conjunction with anatomy for best understanding.

         (d)   all of the above.

(a)   3. The structural and functional unit of life is:

         (a)   a cell.

         (b)   an organ.

         (c)   the organism.

         (d)   a molecule.

(a)   4. Which of the following is in the proper order from small to large?

         (a)   molecules, cells, tissues, organs, organisms

         (b)   cells, molecules, organs, tissues, organ systems

         (c)   cells, molecules, tissues, organs

         (d)   molecules, tissues, cells, organisms

(a, c)   5. Two of the following organ systems bear the major responsibility of ensuring homeostasis of the internal environment. Which two?

         (a)   nervous system

         (b)   digestive system

         (c)   endocrine system

         (d)   reproductive system

         (e)   circulatory system

(e)   6. Hematopoiesis is a function of the:

         (a)   nervous system.

         (b)   endocrine system.

         (c)   cardiovascular system.

         (d)   integumentary system.

         (e)   skeletal system.

7. Match the organs in Column B with the organ systems in Column A.

Column A
(d, j) Nervous
(b) Urinary
(c, f) Circulatory
(e) Digestive
(i) Endocrine
(g) Skeletal
(h) Muscular
(a) Respiratory

Column B
(a) lungs
(b) kidneys
(c) lymph vessels
(d) sense organs
(e) small intestine
(f) heart
(g) skull
(h) voluntary muscles
(i) thyroid gland
(j) brain

(e) 8. Which of the following is a major functional characteristic of all organisms?

(a) movement
(b) growth
(c) metabolism
(d) responsiveness
(e) all of the above

(d) 9. Which of the following is NOT an absolute requirement for life?

(a) oxygen
(b) water
(c) food
(d) atmospheric pressure
(e) adequate body temperature

(a) 10. The term homeostasis means:

(a) maintaining constant conditions within narrow limits.
(b) energy release.
(c) reproduction.
(d) absolutely unchanging.

(d) 11. The elements of physiological control systems include:

(a) a receptor or some kind of sensor.
(b) an effector.
(c) a control center that determines output.
(d) all of the above.

(c) 12. Negative feedback systems:

(a) regulate blood clotting and the birth process.
(b) operate in such a way that the initial stimulus is enhanced and grows.
(c) operate in such a way that the initial stimulus is shut off or reduced.
(d) always inhibit a process as reflected in their naming as "negative."

(c)   13. The glossopharyngeal nerve carries information from blood pressure receptors to
        control centers in the brain stem. This nerve represents a(n):
        (a)   efferent pathway.
        (b)   effector.
        (c)   afferent pathway.
        (d)   variable.

(c)   14. If you were to assume the anatomical position, you would:
        (a)   lie face down.
        (b)   lie flat on your back with the palms of your hands flat on the floor.
        (c)   stand erect with the palms of your hands facing forward.
        (d)   stand erect with the palms of your hands cupped toward your thighs.
        (e)   stand on your head.

(c)   15. The term "sagittal section of the brain" means that the brain was cut:
        (a)   across horizontally.
        (b)   into anterior and posterior parts.
        (c)   into right and left parts.
        (d)   on the diagonal.

(d)   16. Which type of section could be used to separate the thoracic cavity from the
        abdominopelvic cavity?
        (a)   coronal (frontal)
        (b)   sagittal
        (c)   dorsal
        (d)   transverse (horizontal)
        (e)   ventral

(c)   17. Which ventral cavity subdivision has no bony protection?
        (a)   thoracic
        (b)   pelvic
        (c)   abdominal

        18. Relate each of the following statements or conditions to either the dorsal body
        cavity (d) or the ventral body cavity (v).
        (d) Surrounded by the bony skull and vertebral column
        (v) Includes the thoracic and abdominopelvic cavities
        (d) Contains the spinal cord
        (v) Located more anteriorly
        (v) Contains the heart, lungs, and digestive organs

19. Match the key terms with the body cavity locations that follow.

Key: (a) cranial   (d) spinal
     (b) abdominopelvic   (e) none of these
     (c) thoracic

(b) Stomach
(e) Testes
(b) Urinary bladder
(c) Heart
(b) Uterus
(c) Lungs
(b) Liver
(e) Eyes
(a) Brain

20. Choose the structure or organ that matches the given directional term:

wrist   Distal: the elbow/the wrist
hipbone   Lateral: the hipbone/the umbilicus
nose   Superior: the nose/the chin
toes   Anterior: the toes/the heel
scalp   Superficial: the scalp/the skull

(b) 21. All of the following are abdominal regions EXCEPT:

(a) umbilical.
(b) deltoid.
(c) hypogastric.
(d) right iliac.
(e) left hypochondriac.
(f) epigastric.

(a) 22. Which of the following abdominal regions is superior to the umbilical region?

(a) epigastric
(b) hypogastric
(c) right iliac (inguinal)
(d) left lumbar

23. Match the regional anatomy terms in Column B with the body areas in Column A:

Column A	Column B
(a) Arm	(a) brachial
(i) Belly button	(b) cervical
(e) Buttock	(c) digital
(h) Chest	(d) femoral
(c) Fingers and toes	(e) gluteal
(g) Knee (anterior aspect)	(f) oral
(f) Mouth	(g) patellar
(b) Neck	(h) thoracic
(d) Thigh	(i) umbilical

24. Write the name of the organ system in the answer blank preceding its description:

<u>(urinary)</u> Rids the body of nitrogen-containing wastes and regulates the composition of blood.

<u>(endocrine)</u> Includes the pancreas, adrenal and thyroid glands.

<u>(nervous)</u> Responds to environmental changes by transmitting electrical impulses.

<u>(integumentary)</u> Provides an external "envelope" that protects underlying organs.

<u>(digestive)</u> Breaks down ingested food into its building blocks.

<u>(respiratory)</u> Removes carbon dioxide from the blood while adding oxygen to the blood.

<u>(muscular)</u> Enables you to grin, frown, run ,and carry books.

<u>(circulatory)</u> Basically a transportation system; includes the heart and blood vessels.

<u>(reproductive)</u> Allows the production of offspring.

2   Basic Chemistry

Answer the following multiple-choice questions by circling the letter of the correct response. For matching questions, write the correct response in the answer blank.

1. Use choices from the key to identify the energy form in use:

   Key: (a) chemical   (c) mechanical
        (b) electrical   (d) radiant

   ___ (b, d) Vision (two types).
   ___ (c) Bending your fingers to make a fist.
   ___ (a) Breaking the bonds of ATP molecules to energize your muscle cells to make that fist.
   ___ (d) Lying under a sun lamp.

(d)   2. All of the following are examples of the four major elements forming the body EXCEPT:

   (a) hydrogen.
   (b) carbon.
   (c) oxygen.
   (d) calcium.
   (e) nitrogen.

(a)   3. The atomic number that identifies each atom as unique is:

   (a) equal to the number of protons the atom contains.
   (b) the sum of its protons and neutrons.
   (c) the sum of all subparticles.
   (d) the average of the mass numbers of its isotopes.

4. Match the key terms with the descriptions that follow:

   Key: (a) iron   (c) calcium
        (b) iodine   (d) sodium and potassium

   ___ (a) A deficiency of this element can be expected to reduce the hemoglobin content of the blood.
   ___ (d) Necessary for proper conduction of nerve impulses.
   ___ (c) Major structural component of bones and teeth; also required for blood clotting.
   ___ (b) Needed for the thyroid gland to make its hormones.

(b)   5. Which set of terms best describes a proton?

   (a) negative charge, massless, in the orbitalis
   (b) positive charge, 1 amu, in the nucleus
   (c) unchanged 1 amu, in the nucleus

(a) 6. The subatomic particles responsible for the chemical behavior of atoms are:
  (a) electrons.
  (b) protons.
  (c) neutrons.
  (d) ions.

(c) 7. Which of the following statements describes gamma radiation?
  (a) Gamma radiation is a particle.
  (b) Gamma raditation has little penetrating power.
  (c) Gamma radiation is electromagnetic energy.
  (d) Gamma radiation is harmless to living cells.

(c, e) 8. Which of the following are molecules of a compound?
  (a) N_2
  (b) C
  (c) NaOH
  (d) S_8
  (e) $C_6H_{12}O_6$

(a) 9. When a pair of electrons is shared between two atoms, the bond formed is called:
  (a) a single covalent bond.
  (b) a double covalent bond.
  (c) an ionic bond.
  (d) a hydrogen bond.

(b) 10. Molecules formed when electrons are shared unequally are:
  (a) salts.
  (b) polar molecules.
  (c) nonpolar molecules.

  11. Identify each reaction as:
  (a) a synthesis reaction.
  (b) a decomposition reaction.
  (c) an exchange reaction.

  (a) $2 Hg + O_2 -----> 2 HgO$
  (c) $HCL + NaOH -----> NaCl + H_2O$

(d) 12. Water is useful in body processes because:
  (a) it is a good solvent.
  (b) it is a chemical reactant.
  (c) it has a high heat capacity.
  (d) all of the above.

(d) 13. Which is NOT true of all molecules?
  (a) composed of one or more atoms chemically united
  (b) ordinarily in motion
  (c) they are the units of compounds
  (d) always the same size regardless of kind

(b) 14. A solution that has a pH of 10 can best be described as:
    (a) acidic.
    (b) basic.
    (c) neutral.
    (d) naive.

15. Match the terms in Column B with the descriptions in Column A:

Column A	Column B
(d) Building blocks of carbohydrates	(a) amino acids
(a) Building blocks of proteins	(b) fatty acids
(b, c) Building blocks of neutral fats	(c) glycerol
(e) Building blocks of nucleic acids	(d) monosaccharides
	(e) nucleotides

(d) 16. A nucleotide consists of each of the following EXCEPT:
    (a) a nitrogen-containing base.
    (b) a sugar molecule.
    (c) a phosphate group.
    (d) glycerol.

(b) 17. Which of the following is always present in proteins and nucleic acids, but is not usually present in lipids or carbohydrates?
    (a) carbon
    (b) nitrogen
    (c) hydrogen
    (d) iron

(a) 18. Which of the following groups of chemicals includes only polysaccharides?
    (a) glycogen, starch
    (b) glycogen, glucose, galactose
    (c) glucose, fructose
    (d) sucrose, maltose

(a) 19. Which of the following statements about RNA is true?
    (a) RNA is single-stranded.
    (b) RNA is composed of cytosine, guanine, adenine, and thyamine.
    (c) RNA is found only in the nucleus of a cell.
    (d) RNA contains deoxyribose.
    (e) RNA is a double helix.

(b) 20. The genetic information is coded in DNA by:
    (a) the regular arrangement of sugar and phosphate molecules in its backbones.
    (b) the sequence of the bases (A, G, C, and T).
    (c) the number of hydrogen bonds connecting the two strands.
    (d) all of the above.

(c)   21. Carbohydrates, proteins, and lipids are built up from their basic building blocks by the:

       (a) addition of a water molecule between each unit.

       (b) addition of a carbon atom.

       (c) removal of a water molecule between every two units.

       (d) none of the above.

(d)   22. Which of the following is an inorganic molecule?

       (a) sucrose

       (b) cholesterol

       (c) collagen

       (d) sodium chloride

(b)   23. The lipid used as the basis of vitamin D, sex hormones, and bile is:

       (a) neutral fats.

       (b) cholesterol.

       (c) phospholipids.

       (d) prostaglandin.

(e)   24. The molecule that contains high energy phosphate bonds is:

       (a) cholesterol.

       (b) DNA.

       (c) sucrose.

       (d) collagen.

       (e) ATP.

(e)   25. Which statement about enzymes is true?

       (a) They are proteins.

       (b) They form complexes with specific substrates (molecules).

       (c) They are biological catalysts.

       (d) They increase the rates of chemical reactions.

       (e) All of the above.

(d)   26. Which of the following contains amino acids?

       (a) glycogen

       (b) hydrochloric acid

       (c) aldosterone

       (d) antibodies

       (e) vitamin A

     27. Give the correct atomic symbol for each element below:

       (Ca) Calcium

       (Na) Sodium

       (C) Carbon

       (H) Hydrogen

       (N) Nitrogen

       (O) Oxygen

       (K) Potassium

       (Fe) Iron

28. True/False. Determine whether each statement is true or false. If it is true, put a T in the answer blank. If it is false, correct the underlined word or phrase and write your correction in the answer blank.

(nucleus) Most of the mass of an atom is concentrated within the orbitals.

(electrons) An atom becomes an ion when it gains or loses protons.

(T) The most active electrons in an atom occupy its valence shell.

(acceptor) A base may be described as a proton donor.

(T) A substance that has the ability to conduct a current in water is called an electrolyte.

(electrons) Atoms are neutral, therefore, they must contain equal numbers of protons and neutrons.

(neutrons) Isotopes differ from each other only in the number of electrons they contain.

(element) Atoms are the smallest particles of a compound.

(T) The lower the pH, the greater the hydrogen ion concentration.

(structural) Collagen and keratin are examples of functional proteins.

(kinetic) Energy that causes an effect on matter is potential energy.

(T) Organic molecules contain carbon.

3 Cells and Tissues

Answer the following multiple-choice questions by circling the letter of the correct response. For matching questions, write the correct response in the answer blank.

(d) 1. The smallest unit capable of life by itself is:
    (a) an organ.
    (b) an organelle.
    (c) tissue.
    (d) a cell.
    (e) a nucleus.

(a, d) 2. The primary lipids found in the plasma membrane are (choose two):
    (a) phospholipids.
    (b) neutral lipids.
    (c) fat-soluble vitamins.
    (d) cholesterol.

(b) 3. A person drinks a six-pack of beer and has to make several trips to the bathroom. This increase in urination reflects an increase in what membrane transport process occurring in the kidneys?
    (a) dialysis
    (b) filtration
    (c) active transport
    (d) osmosis

(d) 4. Osmosis involves:
    (a) a differentially permeable membrane.
    (b) a difference in water concentration.
    (c) diffusion.
    (d) all of the above.

(b) 5. Choose the group of terms that best describes the process of diffusion:
    (a) passive; carriers; up a concentration gradient
    (b) passive; kinetic energy; down a concentration gradient
    (c) active; carriers; down a concentration gradient
    (d) active; kinetic energy; up or against a concentration gradient

(d) 6. The solute-pumping variety of active transport is accomplished by:
    (a) pinocytosis.
    (b) phagocytosis.
    (c) electrical forces in the membrane.
    (d) ATP-energized carriers.

(c)   7. Microvilli are likely to be found in cells specialized for:
     (a)   contraction.
     (b)   protection.
     (c)   absorption.
     (d)   insulation.
     (e)   division.

(b)   8. The process in which fairly large particles are engulfed and brought into the cell is:
     (a)   pinocytosis.
     (b)   phagocytosis.
     (c)   electrical forces in the membrane.
     (d)   shape and position changes of carrier molecules in the plasma membrane.

9. Match the items in Column B with the appropriate function or description in Column A.

Column A	Column B
(f) Energy-generating powerhouse of the cell.	(a) centriole
(b) Assembles and packages materials to be secreted from the cell.	(b) Golgi apparatus
	(c) inclusions
(e) Provides supports for maintaining cell shape.	(d) lysosome
(l) Synthesizes steroid hormones.	(e) microtubules
(a) Forms the mitotic spindle and the base of cilia.	(f) mitochondria
(j) Site of protein synthesis.	(g) nucleolus
(g) Site of ribosome synthesis.	(h) nucleus
(k) Membrane network studded with ribosomes.	(i) plasma membrane
(h) Control center of the cell.	(j) ribosome
(d) Sac of digestive enzymes.	(k) rough ER
(c) Pigment granules, water vacuoles, etc.	(l) smooth ER

(a)   10. Which of the following organelles consume O_2 (oxygen)?
     (a)   peroxisomes and mitochondria
     (b)   lysosomes and ribosomes
     (c)   lysosomes and peroxisomes
     (d)   mitochondria and rough ER
     (e)   golgi aparatus and centrioles

11. Match the key terms with the stages of mitosis that follow:

    Key: (a) anaphase   (c) prophase
         (b) metaphase (d) telophase

    (c) Centrioles reach the opposite ends of the cell and the chromosomes attach to the spindle.
    (d) Two nuclear membranes are formed.
    (a) The centromeres divide and the chromosomes begin to move apart.
    (b) The chromosomes are precisely lined up on the spindle equator.

(b) 12. Which of the following is not part of the cytoskeleton?

    (a) microtubules
    (b) desmosomes
    (c) intermediate filaments
    (d) microfilaments

(b) 13. The coded information that dictates the structure of a protein molecule is:

    (a) a triplet.
    (b) a gene.
    (c) a nucleotide.

(d) 14. If DNA is a sequence of AAA, then a segment of messenger RNA synthesized on it will have a base sequence of:

    (a) TTT.
    (b) GGG.
    (c) CCC.
    (d) UUU.
    (e) AAA.

(e) 15. Proteins in the plasma membrane:

    (a) may serve as carriers.
    (b) may act in cell-to-cell recognition.
    (c) may help to form pores.
    (d) may serve as hormone receptors.
    (e) all of the above.

(a) 16. Small molecules can pass between adjacent cells if they are connected by:

    (a) gap junctions.
    (b) tight junctions.
    (c) microvilli.
    (d) desmosomes.
    (e) microtubules.

(b) 17. Organized groups of cells with similar structures and a common purpose form a(n):

    (a) organ.
    (b) tissue.
    (c) organism.
    (d) organ system.

18. Match the key terms to the tissue descriptions below:

Key: (a) connective tissue (c) muscle
    (b) epithelium (d) nervous tissue

<u>(b)</u> Filtration, secretion, and absorption are important functions of this tissue.

<u>(a)</u> Composed largely of nonliving extracellular matrix; important in protection and support.

<u>(d)</u> Enables us to be aware of the external environment and to react to it; specialized for communication.

<u>(b)</u> Lines body cavities and covers surfaces; commonly has one surface attached and the other free.

<u>(c)</u> Responsible for essentially all examples of body movements.

(b, c) 19. An epithelium has several layers. The top layer is composed of flattened cells, whereas the basal layer is columnar cells. Choose all terms that would be used in naming this tissue:

(a) simple epithelium.
(b) stratified epithelium.
(c) squamous.
(d) ciliated.
(e) columnar.

20. Identify each gland type as (a) an endocrine gland or (b) an exocrine gland:

(a) ductless
(b) secretes a product through a duct to a membrane surface
(a) products are all hormones
(b) products include milk, sweat, and bile

(a) 21. Which of the following might be part of a cell's extracellular matrix?

(a) interstitial fluid
(b) an inclusion
(c) chromatin
(d) cytosol
(e) lysosomes

22. Match the connective tissue types in Column B with their descriptions in Column A.

Column A

(c) Achilles was "done in" by damage to the tendon connecting his calf muscles to his heel. All tendons consist of this tissue.

(a) No one is literally a "fathead" because the brain is unable to store this tissue.

(g) Has a high content of hard calcium salts.

(b) A soft packing tissue with a soft fluid matrix.

(e) Forms the shock-absorbing pads between the vertebrae

(f) Glassy semihard tissue that covers bone ends at joint surfaces; connects ribs to breastbone.

Column B

(a) adipose
(b) areolar
(c) dense fibrous
(d) elastic cartilage
(e) fibrocartilage
(f) hyaline cartilage
(g) osseous (bone)

(c) 23. The muscle type forming the muscular heart is:

(a) skeletal muscle.
(b) smooth muscle.
(c) cardiac muscle.

24. True/False. Determine whether each statement is true or false. If it is true, put a T in the answer blank. If it is false, correct the underlined word or phrase and write your correction in the answer blank.

(just or exactly) Each daughter cell resulting from mitotic division has half as many chromosomes as the mother cell.

(T) The molecular messengers and decoders of DNA or the genetic material are the RNA molecules.

(chromosomes) The coiled form of the DNA-containing nuclear material is called chromatin.

(interphase) DNA is replicated during early prophase.

(atrophy) A wasting away of a body part because of nonuse or loss of its nerve supply is called hypertrophy.

(solute pumping) Amino acids and most ions are transported through the plasma membrane by faciliated diffusion.

(T) All cells contain a nucleus at least at some period of their lifespan.

(T) A red blood cell placed in a hypertonic solution will lose water.

(nervous) Muscle tissue becomes amitotic shortly after birth and thus is seriously handicapped by severe injury.

(regeneration) Tissue repair by fibrosis or scar tissue formation is the most desirable method.

(metastatic) Another name for a benign neoplasm is cancer.

(cartilage) Although most connective tissues are fairly well supplied with blood vessels, bone is avascular.

(T) Chemicals and physical factors, such as X-rays and UV radiation, are believed to accelerate the aging process.

4   Skin and Body Membranes

Answer the following multiple-choice questions by circling the letter of the correct response. For matching questions, write the correct response in the answer blank.

(c)   1. Where are mucous membranes located?

    (a) in joint cavities
    (b) covering the heart
    (c) lining the stomach
    (d) covering the brain

(b)   2. The most external skin region is composed of:

    (a) simple columnar epithelium.
    (b) stratified squamous epithelium.
    (c) areolar connective tissue.
    (d) dense fibrous connective tissue.

(c)   3. Epidermal cells that are actively mitotic and replace the superficial cells that continually rub off are:

    (a) stratum corneum cells.
    (b) stratum granulosum cells.
    (c) stratum germinativum cells.
    (d) stratum lucidum cells.

(d)   4. Skin color is determined by:

    (a) the amount of carotene in the stratum corneum and subcutaneous tissue.
    (b) pigments in the epidermis (melanin).
    (c) the degree of oxygenation of the blood.
    (d) all of the above.

(d)   5. The dermis:

    (a) when tanned is called leather.
    (b) is composed of connective tissue.
    (c) contains numerous phagocytes (macrophages).
    (d) all of the above.

(d)   6. Which of the following is a function of the skin?

    (a) body temperature control via sweating and heat radiation
    (b) production of vitamin D
    (c) protection from bacterial invasion
    (d) all of the above

(d)   7. Which of the following is NOT an epidermal derivative?

    (a) sweat glands
    (b) sebaceous glands
    (c) hair follicles
    (d) sensory receptors

(a)   8. When the body temperature rises, which of the following takes place?

           (a)   eccrine glands become active
           (b)   arrector pili muscles contract
           (c)   Pacinian corpuscles are stimulated
           (d)   blood is prevented from reaching skin capillary beds
           (e)   all of the above

(b)   9. You can cut hair without feeling pain because:

           (a)   there are no nerve endings associated with hair.
           (b)   the shaft of the hair consists of dead cells.
           (c)   hair follicles are avascular.

(d)   10. Sebum:

           (a)   lubricates the surface of the skin and hair.
           (b)   consists of dead cells and fatty substances.
           (c)   is actively produced when the sebaceous glands are activated by male sex hormones.
           (d)   all of the above.

(a)   11. Which is NOT a true statement about the papillary layer of the dermis?

           (a)   It is the deepest layer of the skin.
           (b)   It produces the pattern for fingerprints.
           (c)   It contains nerve endings that respond to touch and temperature stimuli.
           (d)   It is highly vascular.

(a)   12. The protein found in large amounts in the stratum corneum cells, nails, and hair is:

           (a)   keratin.
           (b)   collagen.
           (c)   melanin.
           (d)   elastin.

(e)   13. The appearance of the skin is helpful in diagnosing certain pathological conditions. For example, it may hint at:

           (a)   anemia.
           (b)   liver malfunction.
           (c)   allergy.
           (d)   respiratory disorder.
           (e)   all of the above.

(b)   14. Joe just burned himself on a hot pot. A blister forms and the burn is painful. Joe's burn is most likely a:

           (a)   first-degree burn.
           (b)   second-degree burn.
           (c)   third-degree burn.
           (d)   Simon Legree burn.

(b)   15. Acne is a disorder associated with inflammation of the:
       (a)   sweat glands.
       (b)   sebaceous glands.
       (c)   hair follicles.

(e)   16. The apocrine sweat glands are:
       (a)   most responsible for odor in humans.
       (b)   found primarily in the axillary and anogenital areas.
       (c)   more often than not associated with hair follicles.
       (d)   become active at puberty.
       (e)   all of the above.

(d)   17. Exposure to the sun:
       (a)   stimulates an increase in the amount of melanin produced.
       (b)   causes the skin to become darker.
       (c)   may cause an increase in the number of melanocytes.
       (d)   a and b are correct.
       (e)   b and c are correct.

(b)   18. Destruction of the hair bulb would result in its inability to:
       (a)   produce gooseflesh.
       (b)   grow.
       (c)   resist drying.
       (d)   produce oil.

(b)   19. The rule of nines is helpful clinically in:
       (a)   diagnosing skin cancer.
       (b)   estimating the extent of a burn.
       (c)   preventing acne.
       (d)   none of the above.

(d)   20. Possible consequences of severe burns include:
       (a)   overwhelming infection.
       (b)   an inability to take in adequate nutrition for repair by eating.
       (c)   circulatory collapse due to tremendous fluid loss.
       (d)   all of the above.

(a)   21. The membrane lining of body cavities open to the exterior is:
       (a)   a mucosa.
       (b)   a serosa.
       (c)   the cutaneous membrane.
       (d)   a synovial membrane.

(d)   22. The membrane composed of connective tissue only is:
       (a)   a mucosa.
       (b)   a serosa.
       (c)   the cutaneous membrane.
       (d)   a synovial membrane.

(b)   23. Which of the following homeostatic imbalanes of the skin is caused by a virus?
   (a)   athlete's foot
   (b)   cold sores
   (c)   impetigo
   (d)   contact dermatitis
   (e)   all of the above

(b)   24. Which of the following relationships is INCORRECT?
   (a)   visceral peritoneum/outer surface of the small intestine
   (b)   parietal pericardium/outer surface of the heart
   (c)   parietal pleura/wall of thoracic cavity

5   The Skeletal System

Answer the following multiple-choice questions by circling the letter of the correct response. For matching questions, write the correct response in the answer blank.

(e)   1. Which is a function of the skeletal system?

    (a) support

    (b) hemopoietic site

    (c) storage

    (d) provides levers for muscle activity

    (e) all of the above

(b)   2. A bone that has essentially the same width, length, and height is most likely:

    (a) a long bone.

    (b) a short bone.

    (c) a flat bone.

    (d) an irregular bone.

(c)   3. The shaft of a long bone is properly called the:

    (a) epiphysis.

    (b) periosteum.

    (c) diaphysis.

    (d) compact bone.

(d)   4. Sites of hematopoiesis include ALL BUT:

    (a) red marrow cavities of spongy bone.

    (b) the diploe of flat bones.

    (c) cavities in bones of infants.

    (d) yellow marrow cavities in bones of a healthy adult.

(d)   5. The osteon exhibits:

    (a) a central Haversian canal carrying a blood vessel.

    (b) concentric lamellae of matrix.

    (c) osteocytes in lacunae.

    (d) canaliculi that connect lacunae to the Haversian canal.

    (e) all of the above.

(c)   6. The organic portion of matrix is important in providing ALL BUT:

    (a) tensile strength.

    (b) hardness.

    (c) ability to resist stretch.

    (d) flexibility.

(c)   7. A young boy fractured his femur in the region of the epiphyseal plate. Which of the following might occur?

      (a) lack of hematopoiesis
      (b) bursitis
      (c) impaired growth of the femur
      (d) rickets

(c)   8. In adult long bones, hyaline cartilage is found:

      (a) in the medullary cavity.
      (b) between trabeculae.
      (c) on articular surfaces.
      (d) in the epiphyseal line.
      (e) in the diaphysis.

(e)   9. Bone growth during childhood and in adults is regulated and directed by:

      (a) growth hormones.
      (b) thyroxine.
      (c) sex hormones.
      (d) mechanical stress.
      (e) all of the above.

(b)   10. A fracture in which the bone ends penetrate soft tissue is:

      (a) greenstick.
      (b) compound.
      (c) simple.
      (d) comminuted.
      (e) compression.

(d)   11. Destruction and remodeling of bone may occur:

      (a) as bones change shape during growth.
      (b) as bones are subjected to varying patterns of stress.
      (c) during and following the healing of a fracture.
      (d) all of the above.

(e)   12. In the healing of bone fractures:

      (a) a blood clot (hematoma) forms around the break.
      (b) granulation tissue forms.
      (c) a fibrocartilage callus forms as a result of fibroblast activity.
      (d) a bony callus eventually replaces the fibrocartilage callus.
      (e) all of the above.

(b)   13. Cranial bones and other flat bones develop:

      (a) from cartilage models.
      (b) on fibrous membranes.
      (c) from a tendon.
      (d) after birth.

(c) 14. What kind of tissue is the forerunner of long bones in the embryo?

 (a) elastic connective tissue
 (b) dense fibrous connective tissue
 (c) hyaline cartilage
 (d) fibrocartilage

(e) 15. PTH:

 (a) is secreted by the parathyroid gland.
 (b) causes activation of the osteoclasts.
 (c) causes increased calcium reabsorption by the kidney tubules.
 (d) is released in response to low blood levels of calcium ions.
 (e) all of the above.

16. Four events are described below. Note in each case whether the immediate controlling factor is: (a) the hormonal control loop mediated by PTH and calcitonin OR (b) the effect of mechanical forces acting on the skeleton.

 (b) A large tubercle develops at the site of origin of a much-used heavy muscle.
 (a) Bone begins to be resorbed as a result of low blood calcium.
 (b) A bedridden patient's bones start to decalcify and atrophy.
 (a) The person is hypercalcemic; bone formation occurs.

17. Match the bone descriptions in Column A with the terms in Column B. (More than one choice may apply.)

Column A	Column B
(b, g) Connected by the frontal suture	(a) ethmoid
(h) Keystone bone of the cranium	(b) frontal
(d) Keystone bone of the face	(c) mandible
(d, f) Form the hard palate	(d) maxillary
(e) Allows the spinal cord to pass	(e) occipital
(c) Forms the chin	(f) palatine
(a, b, d, h) Contain paranasal sinuses	(g) parietal
(i) Contains mastoid sinuses	(h) sphenoid
	(i) temporal

18. Match the key terms with the bone descriptions that follow.

Key: (a) clavicle   (e) sacrum
     (b) ilium   (f) scapula
     (c) ischium   (g) sternum
     (d) pubis   (h) vertebra

 (g) Bone of the axial skeleton to which the pectoral girdle attaches.
 (f) Markings include the glenoid fossa and acromion process.
 (b) Features include the ala, crest, and greater sciatic notch.
 (a) Doubly curved; acts as a shoulder strut to hold the arm away from the thorax.
 (b) Pelvic girdle bone that articulates with the axial skeleton.
 (c) The "sit-down" bone.
 (d) Anteriormost bone of the pelvic girdle.
 (h) Individual bone of the spine.
 (g) Anterior attachment site of the ribs.

19. Indicate whether the following bones are part of the axial (AX) or appendicular (AP) skeleton.

(AX) frontal, (AP) talus, (AX) lumbar vertebra, (AX) sacrum, (AP) radius, (AX) hyoid, (AX) sternum, (AP) clavicle

(a) 20. The spinal cord passes through a large opening in the occipital bone. This opening is an example of a:

(a) foramen.
(b) sinus.
(c) ramus.
(d) facet.
(e) tubercle.

21. Match the key terms with the bone descriptions that follow.

Key: (a) carpals   (e) radius
     (b) femur   (f) tarsals
     (c) fibula   (g) tibia
     (d) humerus   (h) ulna

(b) Articulates with the acetabulum and the tibia.
(c) Forms the lateral aspect of the ankle.
(e) Bone that "carries" the hands.
(a) The wrist bones.
(h) Has a head like a pliers.
(e) Articulates with the capitulum of the humerus.

(d) 22. Which of the following bones is unpaired?

(a) parietal
(b) zygomatic
(c) nasal
(d) mandible
(e) temporal

(b) 23. The inferiormost part of the sternum is the:

(a) body.
(b) xiphoid process.
(c) manubrium.
(d) acromion.

(b) 24. The floating ribs are so named because they have:

(a) no posterior attachment.
(b) no anterior attachment.
(c) no attachment at either end to the axial skeleton.

(c) 25. Which of the following statements regarding the male pelvis is NOT true?

(a) The bones are heavier and rougher than in the female.
(b) The male pelvis is narrow and deep.
(c) The male pubic arch is wider than that in the female.
(d) The cavity of the true pelvis is small (comparatively) in the male.

(d) 26. Since at birth the bones of the cranium have not completed their ossification, there are membranous portions between them called:

  (a) areolas.
  (b) foramina.
  (c) lacunae.
  (d) fontanels.

(c) 27. Which of the following bones is not part of the knee joint?

  (a) femur
  (b) tibia
  (c) fibula
  (d) patella

28. Match the key terms to the appropriate descriptions that follow. (More than one choice may apply.)

Key: (a) synarthroses
       (b) amphiarthroses
       (c) diarthroses (synovial joints)

  (c) Exhibit a joint cavity
  (a) Sutures
  (a) Bones connected by collagen fibers
  (b) Types include intervertebral joints
  (c) All are freely movable
  (b) Slightly movable
  (b) Bones connected by a disk of cartilage
  (a) Immovable

(d) 29. Anatomical characteristics of a synovial joint include:

  (a) articular cartilage.
  (b) a joint cavity.
  (c) an articular capsule.
  (d) all of the above.

(c) 30. Multiaxial joints of the body include:

  (a) the knee and elbow.
  (b) the ankle and wrist.
  (c) the hip and shoulder.
  (d) intercarpal and intertarsal joints.

(c) 31. Rickets most usually develops from a deficiency of which of the following substances?

  (a) vitamin B
  (b) vitamin C
  (c) vitamin D
  (d) calcium

32. Match the key terms with the pathological conditions that follow.

Key: (a) bursitis   (c) osteoarthritis
     (b) gout   (d) rheumatoid arthritis

(c) A consequence of "wear and tear" on joints; chiefly affects the large weight-bearing joints; involves the erosion of articular-cartilage and formation of bony spurs.

(a) Examples are housemaid's knee and tennis elbow.

(b) A painful condition reflecting elevated levels of uric acid in the blood; few joints affected.

(d) An autoimmune disorder; joints affected bilaterally; involves pannus formation and gradual joint immobilization.

33. True/False. Determine whether each statement is true or false. If it is true, put a T in the answer blank. If it is false, correct the underlined word or phrase and write your correction in the answer blank.

(radius) The lateral forearm bone is the ulna.

(T) The ribs articulate with the thoracic vertebrae.

(intervertebral disks) Costal cartilages separate and cushion adjoining vertebrae.

(T) The vomer forms part of the nasal septum.

(T) The lacrimal bone contains openings that allow for tear ducts.

(primary) The thoracic and sacral curvatures of the vertebral column are secondary curvatures.

(T) In women of childbearing age, the dimensions of the false pelvis are of utmost importance.

(acromion) The "point of the shoulder" is formed by the coracoid process of the scapula.

(decreased) The health of bone is increased by a state of weightlessness.

(plate) The structure that promotes growth in the length of a long bone is the epiphyseal line.

(T) Bone forming cells are called osteoblasts.

34. Fill in the missing words to complete the narrative.

A group of high school seniors suffered a serious traffic accident on the way to the prom. There was a pronounced swelling of the upper right side of Mary's head. X-ray films showed a fracture of the largest skull bone, the (parietal bone). Mary also suffered an injury to one of the two large bones of the pelvic girdle. This bone articulates with the sacrum and is named the (ilium). John suffered multiple injuries to his lower-left extremity. Protruding through the skin was a splintered portion of the longest bone in the body, the (femur). Susan didn't consider her injuries serious, so she walked several blocks to find help. Then she noticed that her right knee was not functioning normally. Examination revealed a fractured kneecap. Another name for the kneecap is the (patella).

6   The Muscular System

Answer the following multiple-choice questions by circling the letter of the correct response. For matching questions, write the correct response in the answer blank.

(a)   1. Which of the following groupings is INCORRECT?

   (a) striated muscle, cylindrical cells, attached to the skeleton, involuntary
   (b) smooth muscle, single nucleus, walls of hollow organs, involuntary
   (c) striated muscle, intercalated disks, branching cells, heart
   (d) striated muscle, multinucleate, attached to skeleton, voluntary

(a)   2. Which of the following contains smooth muscle?

   (a) biceps brachii muscle
   (b) heart wall
   (c) stomach wall
   (d) respiratory passages
   (e) both c and d

(b)   3. Which of these surrounds the individual muscle fiber?

   (a) perimysium
   (b) endomysium
   (c) epimysium
   (d) fascia

(d)   4. Which of the following groups of terms is placed in order from largest to smallest?

   (a) myofilament, myofibril, fiber, fascicle
   (b) fiber, myofibril, myofilament, fascicle
   (c) fascicle, fiber, myofilament, myofibril
   (d) fascicle, fiber, myofibril, myofilament
   (e) fiber, myofibril, fascicle, myofilament

(a)   5. The primary function of skeletal muscle is:

   (a) movement of body extremities and locomotion.
   (b) movement of contents through the viscera.
   (c) production of heat.
   (d) ensuring the operation of the heart.

(a)   6. When the neurotransmitter is released and attaches to the receptors on the sarcolemma, _____ rushes into the cell causing _____, which subsequently leads to the contraction process. Choose the correct pair of terms.

   (a) Na^+, depolarization
   (b) Na^+, repolarization
   (c) K^+, depolarization
   (d) K^+, polarization

(d)   7. Complete restoration of the muscle cell to its resting state requires:
        (a) repolarization.
        (b) activation of the sodium-potassium pump.
        (c) depolarization.
        (d) a and b
        (e) b and c

(e)   8. Which of the following areas of the sarcomere become shorter during contraction?
        (a) A band
        (b) I band
        (c) H zone
        (d) a and b
        (e) b and c

(d)   9. Acetylcholine is:
        (a) an ion pump.
        (b) a source of energy for muscle contraction.
        (c) a component of thick myofilaments.
        (d) a neurotransmitter.
        (e) an oxygen binding protein.

(c)   10. The mechanical force of contraction (shortening) is generated by:
        (a) shortening of the thick filaments.
        (b) shortening of the thin filaments.
        (c) the sliding of thin filaments past thick ones.

(a)   11. The sites where the motor nerve impulse is transmitted from the nerve endings to the skeletal muscle cell membranes are the:
        (a) neuromuscular junctions.
        (b) sarcomeres.
        (c) myofilaments.
        (d) Z lines.

(c)   12. A motor unit consists of:
        (a) a skeletal muscle and all of the neurons that supply it.
        (b) a nerve and all of the skeletal muscles supplied by it.
        (c) a neuron and all of the skeletal muscle fibers that it stimulates.
        (d) all of the skeletal muscles that accomplish a single movement.

        13. Use the key terms to complete the statements below.
        Key: (a) actin   (d) A band
             (b) myosin   (e) I band
             (c) Z line

        (c) A sarcomere is the distance between two _____.
        (d) The myosin filaments are located in the _____.
        (e) The _____ contains the actin filaments.
        (b) The thicker filaments are the _____ filaments.

(a)   14. A single, brief, jerky contraction is called:

    (a)   a twitch.
    (b)   wave summation.
    (c)   multiple motor unit summation.
    (d)   tetanus.

(a)   15. The ions that are released inside the muscle cell during action potential generation are:

    (a)   calcium ions.
    (b)   chloride ions.
    (c)   sodium ions.
    (d)   potassium ions.

(e)   16. Tetanus:

    (a)   is a smooth sustained contraction.
    (b)   is a contraction in which the muscle length does not change.
    (c)   is produced by a rapid rate of stimulation.
    (d)   a and b
    (e)   a and c

(a)   17. The proper definition of muscle tone is:

    (a)   a state of sustained partial muscular contraction that is vital for the health of muscles.
    (b)   the feeling of well-being following exercise.
    (c)   the ability of a muscle to maintain a contraction against an outside force.
    (d)   the condition of athletes after intensive training.

(a)   18. Characteristics of isometric contractions include all of the following EXCEPT

    (a)   shortening.
    (b)   increased muscle tension throughout.
    (c)   absence of shortening.
    (d)   use in resistance training.

(c)   19. Aerobic exercise is desirable because it results in all of the following EXCEPT:

    (a)   increased cardiovascular system efficiency.
    (b)   increase in the number of mitochondria in the muscle cells.
    (c)   increase in the size and strength of existing muscle cells.
    (d)   increased neuromuscular system coordination and increased metabolic rate.

(d)   20. Which process provides energy during periods of intense activity?

    (a)   creatine phosphate formation
    (b)   phosphorylation
    (c)   aerobic respiration
    (d)   anaerobic glycolysis

(b) 21. A muscle's least movable point of skeletal attachment is its _____; the more movable end is called its _____. If several muscles have attachments on a particular bone, they can produce complex motions by acting together, or in opposition by acting as _____ to one another. Choose the correct terms:

   (a) origin, insertion, synergists
   (b) origin, insertion, antagonists
   (c) ligament, insertion, synergists
   (d) perimysium, ligament, synergists

22. Match the body movement terms in Column B with the appropriate description in Column A. (More than one choice may apply.)

Column A	Column B
(g) Movement that decreases the angle between two bones; bending the knee.	(a) abduction
	(b) adduction
	(c) circumduction
(a) Movement along the frontal plane, away from the body midline; raising the arm laterally.	(d) dorsiflexion
	(e) eversion
	(f) extension
	(g) flexion
(k) Circular movement around the longitudinal bone axis; shaking the head "no."	(h) inversion
	(i) plantar flexion
	(j) pronation
(c) Describing a cone-shaped pathway with the arm.	(k) rotation
	(l) supination
(l) Moving the hand into a palm-up (or forward) position.	
(d) Movement of the superior aspect of the foot toward the leg; standing on one's heels.	
(h) Turning the sole of the foot medially.	

(a) 23. Paralysis of which of the following would make an individual unable to flex her leg?

   (a) gastrocnemius and hamstrings
   (b) gluteal muscles
   (c) brachioradialis
   (d) soleus

(c) 24. A muscle that assists an agonist by causing a like movement or by stabilizing a joint over which an agonist acts is:

   (a) an antagonist.
   (b) a prime mover.
   (c) a synergist.
   (d) a fixator.

25. Match the muscle names in Column B with the facial muscles described in Column A.

Column A	Column B
(b) Squints the eyes	(a) frontalis
(a) Furrows the forehead horizontally	(b) orbicularis oculi
(d) Smiling muscle	(c) orbicularis oris
(c) Puckers the lips	(d) zygomaticus

(a, c) 26. The prime mover(s) of inspiration is (are) the:

    (a) diaphragm.

    (b) internal intercostals.

    (c) external intercostals.

    (d) abdominal wall muscles.

(c) 27. The arm muscle that both flexes the elbow and supinates the forearm is the:

    (a) brachialis.

    (b) brachioradialis.

    (c) biceps brachii.

    (d) triceps brachii.

(d) 28. The muscles of mastication include:

    (a) buccinators.

    (b) masseters.

    (c) temporalis.

    (d) all of the above.

(b) 29. A prime mover of hip flexion is the:

    (a) rectus femoris.

    (b) iliopsoas.

    (c) vasti muscles.

    (d) gluteus maximus.

(a) 30. The prime mover of hip extension against resistance is the:

    (a) gluteus maximus.

    (b) gluteus medius.

    (c) biceps femoris.

    (d) semimembranosus.

(c) 31. Muscles that cause plantar flexion include ALL BUT the:

    (a) gastrocnemius.

    (b) soleus.

    (c) tibialis anterior.

    (d) peroneus muscles.

32. Match the key terms with the descriptions that follow.

(a) gastrocnemius
(b) semimembranosus
(c) triceps brachii
(d) pectoralis major
(e) sternocleidomastoid
(f) latissimus dorsi
(g) soleus
(h) rectus abdominis
(i) external oblique
(j) rectus femoris
(k) transversus abdominis
(l) semitendinosus
(m) iliopsoas
(n) vasti muscles
(o) biceps femoris
(p) trapezius

(a, g) Two muscles referred to as "toe-dancer" muscles.

(b, l, o) Muscles forming the hamstring group.

(j, n) Muscles forming the quadriceps group.

(e) Main muscle pair used in turning the head to say "no."

(f, p) A good back rub improves circulation to the musculature as well as to the skin. Name the two large superficial paired muscles that are massaged during a back rub from the waist upward.

(h; i, k) Two muscles forming part of your natural abdominal girdle.

(f) Broad muscle of the back that inserts on the humerus.

(d) Antagonist of the muscle named immediately above.

(c) Muscle located on posterior surface of the arm; elbow extensor.

7   The Nervous System

Answer the following multiple-choice questions by circling the letter of the correct response. For matching questions, write the correct response in the answer blank.

(a)   1. Which of the following is NOT a STRUCTURAL feature of a neuron?

      (a)   synapse
      (b)   cell body
      (c)   dendrites
      (d)   axon

(c)   2. A neuron that has as its primary function the job or action of connecting other neurons is called:

      (a)   a motor neuron.
      (b)   a sensory neuron.
      (c)   an association neuron.
      (d)   a glial cell.

(b)   3. The substance released at axonal endings is called a:

      (a)   hormone.
      (b)   neurotransmitter.
      (c)   cholinesterase.
      (d)   none of the above.

(c)   4. Collections of nerve cell bodies outside the central nervous system are called:

      (a)   nuclei.
      (b)   nerves.
      (c)   ganglia.
      (d)   all of the above.

(c)   5. The term "central nervous system" refers to the:

      (a)   autonomic nervous system.
      (b)   brain, spinal cord, and peripheral nerves.
      (c)   brain and spinal cord.
      (d)   spinal cord and spinal nerves.

(d)   6. Which of the following is NOT a function of the autonomic nervous system?

      (a)   innervation of visceral organs
      (b)   regulation and control of vital activities
      (c)   innervation of glands
      (d)   conscious control of motor activities

(d)   7. The neuroglia appear to be involved in:

      (a)   nutrition of the neurons.
      (b)   myelination of the axons of the CNS.
      (c)   support of the nerve cells and phagocytosis of debris.
      (d)   all of the above.

(a)   8. Would a substance that decreases membrane permeability to sodium (a) increase OR (b) decrease the probability of generating a nerve impulse?

(b)   9. In what ways does the interior surface of the membrane of a resting (nonconducting) neuron differ from the external surface?
   (a)   positively charged and contains less sodium
   (b)   negatively charged and contains less sodium
   (c)   negatively charged and contains more sodium
   (d)   positively charged and contains more sodium

(d)   10. The outer covering of a nerve is:
   (a)   endoneurium.
   (b)   perineurium.
   (c)   fascia.
   (d)   epineurium.

(b)   11. Structures that detect the amount of stretch (tension) in skeletal muscles, tendons, and joints are:
   (a)   exteroceptors.
   (b)   proprioceptors.
   (c)   bare dendritic endings.
   (d)   Pacinian receptors.

       12. Match the key terms with the descriptions that follow.
   Key: (a) axon   (c) cell body
        (b) axonal terminal   (d) dendrite

       (b) Releases neurotransmitter.
       (c) The metabolic area.
       (d) Conducts impulses toward the nerve cell body.
       (a) Conducts impulses away from the nerve cell body.

(e)   13. Association neurons are:
   (a)   afferent.
   (b)   efferent.
   (c)   unipolar.
   (d)   bipolar.
   (e)   multipolar.

(a)   14. Impulse conduction is fastest in neurons that are:
   (a)   myelinated.
   (b)   unmyelinated.
   (c)   sensory.
   (d)   motor.
   (e)   cerebral.

15. Match the terms in Column B with the definitions in Column A.

Column A

<u>(c)</u> Period of repolarization of the neuron membrane during which it cannot respond to a second stimulus.

<u>(b)</u> State of reversal of the resting potential, due to an inrush of sodium ions.

<u>(d)</u> Period during which potassium ions diffuse out of the neuron to a change in membrane permeability.

<u>(a)</u> Self-propagated transmission of the depolarization wave along the neuron membrane.

<u>(c)</u> Process during which ATP is used to move sodium out of the cell and potassium into the cell; restores the ionic conditions of the resting (polarized) state.

Column B

(a) action potential
(b) depolarization
(c) refractory period
(d) repolarization
(e) sodium-potassium pump

(a)   16. Myelination of the nerve fibers in the peripheral nervous system is the job of the:

     (a)   Schwann cells.
     (b)   neuroglia.
     (c)   neurons.
     (d)   epithelial cells.

(b)   17. The absolute minimum number of neurons that can be involved in a reflex is/are:

     (a)   1
     (b)   2
     (c)   3
     (d)   4

(c)   18. Which of the following would represent the correct sequence of stimulus transfer in a typical reflex arc?

     (a)   receptor, effector, sensory neuron, association neuron, motor neuron
     (b)   sensory neuron, receptor, association neuron, effector, motor neuron
     (c)   receptor, sensory neuron, association neuron, motor neuron, effector
     (d)   receptor, association neuron, sensory neuron, motor neuron

(a)   19. Which of the following terms describes the patellar reflex?

     (a)   somatic
     (b)   visceral
     (c)   autonomic
     (d)   central
     (e)   cranial

(c)   20. A spinal nerve has:
       (a)   motor pathways.
       (b)   sensory fibers.
       (c)   both sensory and motor fibers.

(a)   21. The central fissure lies between the:
       (a)   frontal lobe and parietal lobe.
       (b)   right and left cerebral hemispheres.
       (c)   cerebrum and cerebellum.
       (d)   parietal lobe and temporal lobe.
       (e)   parietal lobe and occipital lobe.

(a, b) 22. A depression on the surface of the cortex is called a (choose two):
       (a)   sulcus.
       (b)   fissure.
       (c)   gyrus.
       (d)   furrow.
       (e)   map.

23. Match the cerebral hemisphere lobes (key terms) with the function areas they contain.
    Key: (a) frontal   (c) occipital
          (b) parietal   (d) temporal

      (d) Auditory area
      (b) Somatic (general) sensory cortex
      (a) Somatic motor cortex
      (a) Broca's (motor speech) area
      (a) Premotor area
      (a) Seat of intelligence, abstract reasoning
      (c) Visual areas
      (d) Language/speech comprehension area
      (b) Taste (gustatory) area

24. Match the key terms for the statements describing various brain areas that follow.

Key: (a) basal nuclei (f) medulla
(b) cerebellum (g) midbrain
(c) corpora quadrigemina (h) pons
(d) corpus callosum (i) thalamus
(e) hypothalamus

(a) Involved in fine control of motor activities; lesions may lead to Parkinson's disease.

(f) Region where there is a gross crossover of fibers of descending pyramidal tracts.

(e) Control of temperature, ANS reflexes, hunger, and water balance.

(c) Relay stations for visual and auditory stimuli input; found in midbrain.

(f) Houses vital centers for control of the heart, respiration, and blood pressure.

(i) Region through which (nearly) all sensory input must travel to get to the cerebral cortex; acts as a sensory relay area.

(a) Region most concerned with equilibrium, body posture, and coordination of motor activity.

(d) 25. The innermost layer of the meninges, delicate and closely apposed to the brain tissue, is the:

(a) dura mater.
(b) corpus callosum.
(c) arachnoid.
(d) pia mater.

(c) 26. Cerebrospinal fluid is formed by:

(a) arachnoid villi.
(b) the dura mater.
(c) choroid plexuses.
(d) all of the above.

(e) 27. Which of the following blood-borne substances can cross capillary walls in the brain?

(a) urea
(b) K+
(c) glucose
(d) respiratory gases
(e) both c and d

(a) 28. A patient has suffered a cerebral hemorrhage that has damaged the precentral gyrus of his right cerebral cortex. As a result:

(a) he cannot voluntarily move his left arm or leg.
(b) he feels no sensation on the left side of his body.
(c) he feels no sensation on the right side of his body.

(b)   29. Ascending pathways in the spinal cord convey:
    (a)   motor impulses.
    (b)   sensory impulses.
    (c)   voluntary impulses.
    (d)   all of the above.

(c)   30. Destruction of the anterior horn cells of the spinal cord results in loss of:
    (a)   integrating impulses.
    (b)   sensory impulses.
    (c)   voluntary motor impulses.
    (d)   all of the above.

(b)   31. Brain waves typical of the alert, wide-awake state are:
    (a)   alpha.
    (b)   beta.
    (c)   delta.
    (d)   theta.

32. Match the names of the cranial nerves in Column B with the descriptions in Column A.

Column A	Column B
(f) Causes pupillary constriction.	(a) abducens
(c) Is the major sensory nerve of the face.	(b) accessory
	(c) facial
(b) Serves the sternocleidomastoid and trapezius muscles.	(d) glossopharyngeal
	(e) hypoglossal
(g, h, l) Are purely sensory (three nerves).	(f) oculomotor
	(g) olfactory
(e) Serves the tongue muscles.	(h) optic
(i) Enables you to chew your food.	(i) trigeminal
(k) Helps regulate heart activity.	(j) trochlear
(l) Serves your ability to hear and maintain your balance.	(k) vagus
	(l) vestibulocochlear

33. For each of the following muscles or body regions, identify the plexus and the peripheral nerve (or branch of one) involved. Use choices from Keys A and B.

Key A: Plexuses
    (a) brachial
    (b) cervical
    (c) lumbar
    (d) sacral

Key B: Nerves
    (1) common peroneal
    (2) femoral
    (3) median
    (4) musculocutaneous
    (5) obturator
    (6) phrenic
    (7) radial
    (8) tibial
    (9) ulnar

<u>(b), (6)</u> The diaphragm
<u>(d), (1, 8)</u> Muscles of the posterior thigh and leg
<u>(c), (2)</u> Anterior thigh muscles
<u>(a),(3, 4, 7, 9)</u> Muscles of the arm and forearm (four nerves)

(c) 34. Which of these effectors is NOT controlled by the autonomic nervous system?
    (a) smooth muscle
    (b) cardiac muscle
    (c) skeletal muscle
    (d) most glands

(d) 35. Cranial nerves containing preganglionic parasympathetic fibers include ALL BUT:
    (a) facial.
    (b) glossopharyngeal.
    (c) oculomotor.
    (d) trigeminal.
    (e) vagus.

(e) 36. Which of the following is NOT a result of parasympathetic stimulation?
    (a) salivation
    (b) constriction of the pupils
    (c) increased peristalsis of the digestive viscera
    (d) relaxation of the bladder sphincter
    (e) none of the above

(e) 37. The parasympathetic division of the autonomic nervous system has NO effect on:
    (a) the liver.
    (b) the kidneys.
    (c) mental activity.
    (d) metabolic rate.
    (e) all of the above.

(d) 38. Cardiovascular effects of the sympathetic division include ALL BUT:
- (a) constriction of most blood vessels.
- (b) dilation of the vessels serving skeletal muscles.
- (c) increase of heart rate and force.
- (d) dilation of the blood vessels serving the skin and digestive viscera.

(c) 39. Most body organs are innervated by:
- (a) the parasympathetic division.
- (b) the sympathetic division.
- (c) both the sympathetic and parasympathetic divisions.

(a) 40. The site of origin of the preganglionic fibers of the sympathetic nervous system is:
- (a) the thoracolumbar region of the spinal cord.
- (b) the higher brain centers.
- (c) the sympathetic chain.
- (d) the brain stem and the sacral region of the cord.

(b) 41. The substance released by the axonal endings of the somatic efferent fibers and by the parasympathetic nerve fibers is:
- (a) norepinephrine.
- (b) acetylcholine.
- (c) serotonin.
- (d) adrenalin.

(e) 42. A patient scheduled for surgery tells his nurse that he is very scared. Which of the following indicate increased sympathetic activity in this patient?
- (a) He complains that his mouth feels very dry.
- (b) His gown is damp with perspiration.
- (c) His skin is pale.
- (d) His heart is pounding.
- (e) All of the above.

(c) 43. The epinephrine and norepinephrine secretions of the adrenal medulla act to supplement the effects of:
- (a) parasympathetic fibers.
- (b) vagus nerve activity.
- (c) sympathetic stimulation.
- (d) somatic nervous system.

(a) 44. In contrast to the somatic nervous system, the autonomic nervous system:
- (a) has two efferent neurons.
- (b) has two afferent neurons.
- (c) stimulates its effector cells.
- (d) has both afferent and efferent fibers.

(a)   45. Preparing the body for the "fight-or-flight" response is the role of the:

    (a)   sympathetic nervous system.

    (b)   cerebrum.

    (c)   parasympathetic nervous system.

    (d)   none of the above.

46. True/False. Determine whether each statement is true or false. If it is true, put a T in the answer blank. If it is false, correct the underlined word or phrase and write your correction in the answer blank.

    (efferent)   Afferent neurons transmit impulses from the periphery to the CNS.

    (rapidly)   Large nerve fibers conduct impulses more slowly than do small fibers.

    (T)   Neurons are amitotic after birth.

    (T)   In myelinated axons the sodium channels are concentrated at the nodes of Ranvier.

    (hypothalamus)   One of the major functions of the pons is to regulate the endocrine system by producing releasing factors that control the function of the anterior pituitary.

    (cerebrum)   Basal (cerebral) nuclei are gray matter areas buried deep within the white matter of the cerebellum.

    (encephalitis)   Meningitis is the most accurate term for inflammation of the brain.

    (subarachnoid)   Cerebrospinal fluid circulates within the ventricles of the brain and in the subdural space outside the brain.

    (L2)   The spinal cord ends at the level of T12.

    (capillary impermeability)   The crucial factor in construction of the blood-brain barrier is the positioning of the astrocytes between the capillaries and the neurons.

    (T)   Spina bifida is an abnormal fusion of the bony vertebral parts that may or may not involve the meninges and spinal cord.

    (thalamus)   The bulk of the diencephalon is formed by the hypothalamus.

    (pons)   Difficulty in breathing may reflect damage to respiratory centers located in the medulla and midbrain.

    (RAS)   The limbic system acts as a sensory filter for the brain and also acts in motor pathways.

    (left)   The dominant cerebral hemisphere is usually the right.

8   Special Senses

Answer the following multiple-choice questions by circling the letter of the correct response. For matching questions, write the correct response in the answer blank.

(a)   1. Which of the following is NOT monitored by a special sense receptor?

     (a)   pain
     (b)   sight
     (c)   hearing
     (d)   smell
     (e)   taste

(c)   2. The accessory glands that produce an oily secretion are the:

     (a)   conjunctiva.
     (b)   lacrimal glands.
     (c)   meibomian glands.

(d)   3. The portion of the fibrous tunic that is white and opaque is the:

     (a)   choroid.
     (b)   cornea.
     (c)   retina.
     (d)   sclera.

(b)   4. Which sequence best describes a normal route for the flow of tears from the eyes into the nasal cavity?

     (a)   lacrimal canals, nasolacrimal ducts, nasal cavity
     (b)   lacrimal ducts, lacrimal sacs, nasolacrimal ducts
     (c)   nasolacrimal ducts, lacrimal canals, lacrimal sacs

(b)   5. Four refractory media of the eye, listed in the sequence in which they refract light, are:

     (a)   vitreous humor, lens, aqueous humor, cornea.
     (b)   cornea, aqueous humor, lens, vitreous humor.
     (c)   cornea, vitreous humor, lens, aqueous humor.
     (d)   lens, aqueous humor, cornea, vitreous humor.

(c)   6. Damage to the medial recti muscles would probably affect:

     (a)   accommodation.
     (b)   refraction.
     (c)   convergence.
     (d)   pupil constriction.

(d)   7. The phenomenon of light adaptation is best explained by the fact that:

     (a)   rhodopsin does not function in dim light.
     (b)   rhodopsin breakdown occurs slowly.
     (c)   rods exposed to intense light need time to generate rhodopsin.
     (d)   cones are stimulated to function by bright light.

(b) 8. Blockage of the canal of Schlemm might result in:

    (a) a sty.
    (b) glaucoma.
    (c) conjunctivitis.
    (d) a cataract.

(a) 9. Nearsightedness is more properly called:

    (a) myopia.
    (b) hyperopia.
    (c) presbyopia.
    (d) emmetropia.

(b) 10. Of the neurons in the retina, the axons of which of these form the optic nerve?

    (a) bipolar neurons
    (b) ganglion cells
    (c) cone cells
    (d) horizontal cells

(b) 11. Which sequence of reactions occurs when a person looks at a distant object?

    (a) Pupils constrict, suspensory ligaments relax, lenses become less convex.
    (b) Pupils dilate, suspensory ligaments become taut, lenses become less convex.
    (c) Pupils dilate, suspensory ligaments become taut, lenses become more convex.
    (d) Pupils constrict, suspensory ligaments relax, lenses become more convex.

(c) 12. The main function of the rods is:

    (a) depth perception.
    (b) color vision.
    (c) vision in dim light.
    (d) accommodation for near vision.
    (e) refraction.

(a) 13. The function of the auditory tube is:

    (a) to equalize air pressure on both sides of the tympanic membrane.
    (b) to equalize blood pressure in all parts of the body during exercise.
    (c) to magnify sounds.
    (d) to help maintain equilibrium.

(c) 14. The cortical region for vision is located in the:

    (a) temporal lobe.
    (b) frontal lobe.
    (c) occipital lobe.
    (d) parietal lobe.

(b) 15. The fovea centralis is:

    (a) also known as the blind spot.
    (b) an area that contains a high concentration of cones.
    (c) an area that contains a high concentration of rods.
    (d) a vascular tunic.
    (e) a rounded opening in the iris.

(a)   16. Presbyopia is usually due to:
  (a) loss of elasticity of the lens.
  (b) unequal curvature of refracting surfaces.
  (c) an eyeball that is too long.
  (d) a flattened cornea.

17. Match the structures in Column B with the definitions in Column A:

Column A	Column B
(e) Particles of calcium carbonate.	(a) crista ampullaris
(c) Connects cavity of middle ear with pharynx.	(b) cochlear duct
	(c) auditory tube
(f) Externally visible fleshy part of the ear.	(d) organ of Corti
	(e) otoliths
(h) Separates external auditory canal from the tympanic cavity.	(f) pinna
(i) Contains the utricle and saccule.	(g) semicircular canals
(b) Divides the cochlea into chambers.	(h) tympanic membrane
(a or g) Contains receptors for dynamic equilibrium.	(i) vestibule
(d) Contains receptors for hearing.	
(i) Part of the bony labyrinth connecting the cochlea and the semicircular canals.	
(h) Communicates sound vibrations to the ossicles, set into motion by sound waves.	

18. Match the key terms with the descriptions that follow.
  Key: (a) conjunctiva
       (b) lacrimal glands
       (c) meibomian glands

  (b) Product is a saline solution containing lysozyme and antibodies.
  (c) Produce oil.
  (a) Secretes mucus.

19. Match the key terms with the descriptions that follow.
  Key: (a) ciliary body   (e) iris
       (b) ciliary processes (f) retina
       (c) choroid   (g) sclera
       (d) cornea

  (f) Portion (tunic) of the eye concerned with image formation.
  (c) Vascular, light-absorbing tunic part.
  (a) Muscular structure that manipulates the lens.
  (g) Supports and protects the eyeball.
  (e) Controls the amount of light entering the eye.

(b)   20. Tinnitus, vertigo, and gradual hearing loss typify the disorder called:

      (a)   conjunctivitis.
      (b)   Ménière's syndrome.
      (c)   strabismus.
      (d)   motion sickness.

(c)   21. Inflammation of the middle ear, commonly seen in children, is called:

      (a)   otosclerosis.
      (b)   anosmia.
      (c)   otitis media.
      (d)   presbycusis.
      (e)   Ménière's syndrome.

(d)   22. Which of the following structures is NOT part of the external ear?

      (a)   pinna
      (b)   external auditory meatus
      (c)   tympanic membrane
      (d)   oval window

(b)   23. Which of the following could NOT be seen as one looks into the eye with an ophthalmoscope?

      (a)   macula lutea
      (b)   optic chiasma
      (c)   fovea centralist
      (d)   optic disk

(e)   24. Which of the following is important in maintaining the balance of the body?

      (a)   visual cues
      (b)   semicircular canals
      (c)   the saccule
      (d)   proprioceptors
      (e)   all of the above

(b)   25. Static equilibrium receptors that report the position of the head in space relative to the pull of gravity are:

      (a)   organ of Corti.
      (b)   maculae.
      (c)   crista ampullaris.

(b)   26. The receptors for hearing in the inner ear are called:

      (a)   crista ampullaris.
      (b)   hair cells.
      (c)   maculae.
      (d)   otoliths.
      (e)   endolymph.

(c)   27. Which of the following is NOT a possible cause of conduction deafness?

(a)   impacted cerumen
(b)   middle ear infection
(c)   cochlear nerve degeneration
(d)   otosclerosis

(d)   28. Olfactory nerve damage would probably affect your ability to:

(a)   taste.
(b)   hear.
(c)   feel pain.
(d)   smell.

(a)   29. Sensory impulses transmitted over the facial, glossopharyngeal, and vagus nerves are involved in the sensation of:

(a)   taste.
(b)   touch.
(c)   equilibrium.
(d)   smell.

(d)   30. Taste buds are found on the:

(a)   anterior part of the tongue.
(b)   posterior part of the tongue.
(c)   palate.
(d)   all of the above.

(c)   31. Gustatory cells are stimulated by:

(a)   the movement of otoliths.
(b)   stretch.
(c)   chemicals in solution.
(d)   light.

(d)   32. The olfactory receptor cells can be characterized by all the following statements EXCEPT:

(a)   They are ciliated.
(b)   They are neurons.
(c)   They are chemoreceptors.
(d)   They are epithelial cells wrapped by sensory nerve endings.
(e)   They show a range of responsiveness to chemicals.

(a)   33. The region of the tongue with the greatest sensitivity to sweet tastes is the:

(a)   tip.
(b)   sides.
(c)   back (posterior aspect).
(d)   all areas are equally sensitive.

(

(e)   34. Part of "taste" derives from stimulation of
   (a)   gustatory cells.
   (b)   olfactory receptors.
   (c)   pain receptors.
   (d)   temperature receptors.
   (e)   all of the above.

35. True/False. Determine whether each statement is true or false. If it is true, put a T in the answer blank. If it is false, correct the underlined word or phrase and write your correction in the answer blank.

(aqueous) The anterior chamber of the eye is filled with <u>vitreous</u> humor.

(cortical) Sensations are interpreted by the <u>receptor cells</u>.

(refraction) The bending of light rays is called <u>reflection</u>.

(damp/moist) <u>Dry</u> air is important in olfactory reception.

(ethmoid) The olfactory nerve fibers pierce the cribriform plate of the <u>sphenoid</u> bone.

(more) Contraction of the ciliary muscle causes the lens to become more convex, thus bending the light <u>less</u>.

(cornea) The transparent structure through which light enters the eye is the <u>pupil</u>.

(low) Sounds of <u>high</u> frequency stimulate the hair cells at the apex of the organ of Corti.

(converge) The function of the lens is to <u>diverge</u> the light rays and focus them on the retina.

(temporal) The lateral wall of the middle ear is formed primarily by the <u>parietal</u> bone.

(endolymph) The fluid that fills the membranous labyrinth is <u>perilymph</u>.

9 The Endocrine System

Answer the following multiple-choice questions by circling the letter of the correct response. For matching questions, write the correct response in the answer blank.

(d)   1. A hormone is a chemical that:
  (a) is produced by living cells.
  (b) regulates metabolic activities.
  (c) is distributed via the blood or lymph.
  (d) all of the above.

(c)   2. Which of the hormones listed next is NOT steroid-based?
  (a) estrogen
  (b) aldosterone
  (c) epinephrine
  (d) testosterone
  (e) cortisol

       3. Indicate the major stimulus (key term) for release of each of the hormones that follow.
       Key: (a) hormonal   (c) neural
            (b) humoral

  (a) Adrenocorticotropic hormone
  (b) Parathyroid hormone
  (b) Insulin
  (a) Thyroxine and triiodothyronine
  (c) Epinephrine

(b)   4. Which of the following elements is required for the synthesis of thyroxine?
  (a) iron
  (b) iodine
  (c) sodium
  (d) potassium

(c)   5. Which of the following statements describes steroid hormones?
  (a) They bind to receptors on the target cell plasma membrane.
  (b) They activate membrane associated enzymes.
  (c) They activate specific target cell genes.
  (d) They lead to the production of second messengers.

(d)   6. Relative to a second messenger system, which of the following is NOT accurate?
  (a) The activating hormone interacts with a receptor site on the plasma membrane.
  (b) The activating hormone activates an enzyme, often adenylate cyclase.
  (c) Activated adenylate cyclase catalyzes the transformation of AMP to cyclic AMP.
  (d) Cyclic AMP directly alters cell function as is characteristic for that specific hormone.

(d)   7. Some hormones act by:

    (a)   increasing the synthesis of enzymes.

    (b)   converting an inactive enzyme into an active enzyme.

    (c)   affecting only specific target organs.

    (d)   all of these.

8. Match the key terms with the descriptions that follow.

Key: (a) aldosterone   (e) oxytocin
     (b) antidiuretic hormone   (f) prolactin
     (c) growth hormone   (g) T_4 and T_3
     (d) luteinizing hormone

    (c) Important anabolic hormone.

    (b) Involved in water balance; causes the kidneys to conserve water.

    (f) Stimulates milk production.

    (d) Tropic hormone that stimulates the gonads to secrete sex hormones.

    (e) Increases uterine contractions during birth.

    (g) Major metabolic hormone(s) of the body.

    (d) Causes reabsorption of sodium ions by the kidneys.

(a)   9. The anterior pituitary does NOT secrete:

    (a)   antidiuretic hormone.

    (b)   growth hormone.

    (c)   gonadotropins.

    (d)   TSH.

(b)   10. Tropic hormones:

    (a)   control growth.

    (b)   stimulate hormone secretion by other endocrine glands.

    (c)   promote water retention.

    (d)   control glucose metabolism.

(c)   11. A hormone NOT involved with sugar metabolism is:

    (a)   glucagon.

    (b)   cortisone.

    (c)   aldosterone.

    (d)   insulin.

(d)   12. Parathyroid hormone:

    (a)   increases bone formation and lowers blood calcium levels.

    (b)   increases calcium excretion from the body.

    (c)   decreases calcium absorption in the intestine.

    (d)   demineralizes bone and raises blood calcium levels.

(d)   13. A hypodermic injection of epinephrine would:

    (a)   increase heart rate, increase blood pressure, dilate the bronchi of the lungs, and increase peristalsis.

    (b)   decrease heart rate, decrease blood pressure, constrict the bronchi, and increase peristalsis.

    (c)   decrease heart rate, increase blood pressure, increase peristalsis, and decrease peristalsis.

    (d)   increase heart rate, increase blood pressure, dilate the bronchi, and decrease peristalsis.

(c)   14. Testosterone is to the male as what hormone is to the female?

    (a)   luteinizing hormone

    (b)   progesterone

    (c)   estrogen

    (d)   prolactin

(b)   15. If anterior pituitary secretion is deficient in a growing child, the child will:

    (a)   develop acromegaly.

    (b)   become a dwarf but have fairly normal body proportions.

    (c)   mature sexually at an earlier-than-normal age.

    (d)   be in constant danger of becoming dehydrated.

(a)   16. The endocrine gland that develops in common with the sympathetic nervous system is the:

    (a)   adrenal medulla.

    (b)   pancreas.

    (c)   thyroid.

    (d)   anterior pituitary gland.

(e)   17. The hormone that helps to reduce blood volume and blood pressure is:

    (a)   renin.

    (b)   aldosterone.

    (c)   epinephrine.

    (d)   antidiuretic hormone.

    (e)   atrial natriuretic factor.

(d)   18. Sexual development is influenced by the:

    (a)   pituitary.

    (b)   gonads.

    (c)   adrenal glands.

    (d)   all of the above.

(e)   19. Symptoms of diabetes mellitus include:

    (a)   polydypsia.

    (b)   polyphagia.

    (c)   polyuria.

    (d)   hyperglycemia.

    (e)   all of the above.

(c) 20. Individuals with Type II diabetes mellitus:

    (a) have an absolute requirement for insulin injections.

    (b) are often in ketosis.

    (c) may often control their disease by diet and exercise.

    (d) have very efficient insulin receptors.

(b) 21. Hypothyroidism in children may lead to:

    (a) Graves' disease.

    (b) cretinism.

    (c) myxedema.

    (d) exophthalmos.

22. True/False. Determine whether each statement is true or false. If it is true, put a T in the answer blank. If it is false, correct the underlined word or phrase and write your correction in the answer blank.

(T) Cortisol enables the body to cope effectively with stressful situations of <u>long</u> duration.

(steroid) Most <u>protein/amino acid</u> hormones bring about their effects by interacting with the genes.

(T) The more active form of thyroid hormone is T_3.

(T) A major effect of <u>insulin</u> is to facilitate the uptake of glucose by many cells.

(negative) Secretion of most hormones is regulated by <u>positive</u> feedback; that is, as hormone levels rise, the stimulus for hormone release ends.

(hypothalamus) The group of hormones called releasing hormones are produced by the <u>pineal gland</u>.

(glucocorticoids) Cushing's syndrome mainly reflects an excess of <u>mineralocorticoids</u>.

(sperm) In males, FSH stimulates the production of <u>testosterone</u>.

(antagonize) PTH and calcitonin <u>enhance</u> each other's effects.

(T) <u>Tropic hormones</u> influence the activity of other hormone-producing organs.

10   Blood

Answer the following multiple-choice questions by circling the letter of the correct response. For matching questions, write the correct response in the answer blank.

(c)   1. The blood volume in an adult averages approximately:
      (a) 1 L.
      (b) 3 L.
      (c) 5 L.
      (d) 7 L.

(b)   2. The formed elements are suspended in an extracellular matrix called:
      (a) serum.
      (b) plasma.
      (c) lymph.
      (d) water.

(c)   3. The hormone that prompts red blood cell formation is:
      (a) serotonin.
      (b) heparin.
      (c) erythropoietin.
      (d) thrombopoietin.

(d)   4. All of the following are true of RBCs EXCEPT:
      (a) biconcave disk shape.
      (b) lifespan of approximately 120 days.
      (c) contain hemoglobin.
      (d) contain nuclei.

(b)   5. The most numerous white blood cell is the:
      (a) eosinophil.
      (b) neutrophil.
      (c) monocyte.
      (d) lymphocyte.

(a)   6. Reticulocytes develop into:
      (a) erythrocytes.
      (b) lymphocytes.
      (c) hemocytoblasts.
      (d) monocytes.
      (e) platelets.

(d)   7. Blood proteins play an important part in:
      (a) blood clotting.
      (b) immunity.
      (c) maintenance of blood volume.
      (d) all of the above.

(a) 8. The white blood cell that releases histamine and heparin is the:
- (a) basophil.
- (b) neutrophil.
- (c) monocyte.
- (d) eosinophil.

(a) 9. The blood cell that is part of the immune system is the:
- (a) lymphocyte.
- (b) megakaryocyte.
- (c) neutrophil.
- (d) basophil.

(b) 10. The normal erythrocyte count (per cubic millimeter) for adults is:
- (a) 3 to 4 million.
- (b) 4.5 to 5 million.
- (c) 8 million.
- (d) 500,000.

(c) 11. The normal pH of the blood is about:
- (a) 8.4.
- (b) 7.8.
- (c) 7.4.
- (d) 4.7.

(d) 12. Suppose your blood was found to be AB positive. This means that:
- (a) antigens A and B are present on your red blood cells.
- (b) there are no anti-A or anti-B antibodies in your plasma.
- (c) your blood is RH^+.
- (d) all of the above.

(d) 13. A sample of type O blood will agglutinate with:
- (a) anti-A serum.
- (b) anti-B serum.
- (c) both anti-A serum and anti-B serum.
- (d) neither anti-A serum or anti-B serum.

(b) 14. All of the following conditions impair coagulation EXCEPT:
- (a) vitamin K deficiency.
- (b) vascular spasms.
- (c) severe hypocalcemia.
- (d) heparin.
- (e) liver disease.

(c) 15. The activator of the reaction fibrinogen ---> fibrin is:
- (a) prothrombin activator.
- (b) factor XIII.
- (c) thrombin.
- (d) thromboplastin.

(d)   16. In the healthy adult, erythrocytes are formed in the:

    (a)   liver.
    (b)   yellow bone marrow.
    (c)   lymphoid tissue.
    (d)   red bone marrow.

(a, d) 17. If an Rh- mother becomes pregnant, when can hemolytic disease of the newborn NOT possibly occur in the child? (Select all that apply.)

    (a)   if the child is Rh-
    (b)   if the child is Rh+
    (c)   if the father is Rh+
    (d)   if the father is Rh-

(a)   18. Which of the following plasma proteins is most important in maintaining the proper osmotic pressure of the blood?

    (a)   albumin
    (b)   gamma globulin
    (c)   fibrinogen
    (d)   beta globulin

(a)   19. The erythrocyte count increases when an individual goes from a low to a high altitude because:

    (a)   the concentration of oxygen and/or total atmospheric pressure is lower at high altitudes.
    (b)   the basal metabolic rate is higher at high altitudes.
    (c)   the concentration of oxygen and/or total atmospheric pressure is higher at high altitudes.
    (d)   the temperature is lower at high altitudes.

(d)   20. Which of the following would NOT characterize leukocytes?

    (a)   ameboid
    (b)   phagocytic (some)
    (c)   nucleated
    (d)   the cells are found in largest numbers in the bloodstream
    (e)   exhibit chemotaxis

(c)   21. Which of the following is an anticoagulant?

    (a)   thromboplastin
    (b)   fibrinogen
    (c)   heparin
    (d)   serotonin
    (e) PF_3

22. Match the terms in Column B with the descriptions in Column A:

Column A

(a) Blood-borne antibodies.

(p) Chemical released by platelets that causes spasms of the blood vessel.

(l) Plasma without the clotting proteins.

(i) Respiratory pigment in RBCs that binds oxygen.

(f) Floating intravascular clot.

(n) Adverse reaction of donor's cells with recipient's plasma.

(j) Classic "bleeder's" disease.

(o) Plasma plus formed elements.

(k) Liquid portion of the blood.

(b) Any decrease in oxygen-carrying ability of the blood.

(c) Transport proteins found in the blood.

(e) Chemicals that prompt WBC production.

(d) Coagulation.

Column B

(a) agglutinins

(b) anemia

(c) beta-globulins

(d) clotting

(e) colony-stimulating factors

(f) embolus

(g) gamma globulins

(h) factor VIII

(i) hemoglobin

(j) hemophilia

(k) plasma

(l) serum

(m) thrombus

(n) transfusion reaction

(o) whole blood

(p) serotonin

(b) 23. Which of the following cells is a precursor of a basophil?

(a) megakaryoblast

(b) myeloblast

(c) lymphoblast

(d) normoblast

(d) 24. Complications of aplastic anemia include:

(a) bleeding disorders.

(b) hypoxia of body tissues.

(c) lethal infection.

(d) all of the above.

(c) 25. The special type of hemoglobin present in fetal red blood cells is:

(a) hemoglobin A.

(b) hemoglobin B.

(c) hemoglobin F.

(d) hemoglobin S.

(a) 26. All of the following can be expected with polycythemia EXCEPT:

(a) low blood viscosity.

(b) high hematocrit.

(c) increased blood volume.

(d) high blood pressure.

(b)   27. Which of the following is NOT a cause of RBC sickling in sickle-cell anemia?
    (a) blood loss
    (b) leukemia
    (c) travel at high altitude
    (d) vigorous exercise

(a)   28. No cytoplasmic granules are present in:
    (a) monocytes.
    (b) basophils.
    (c) eosinophils.
    (d) neutrophils.

(a)   29. James has a hemoglobin measurement of 16 g/100 mL of blood. This is:
    (a) within the normal range.
    (b) low normal.
    (c) abnormally low.
    (d) above normal.

(c)   30. Which of these is NOT a plasma protein?
    (a) fibrinogen
    (b) gamma globulins
    (c) thromboplastin
    (d) beta globulins
    (e) albumin

(c)   31. Together leukocytes and platelets comprise what percentage of whole blood volume?
    (a) 45
    (b) 55
    (c) 1
    (d) 10

(a)   32. Which is the correct sequence of the following events?
    1. Fibrinogen ---> fibrin
    2. Clot retraction
    3. Release of thromboplastin
    4. Prothrombin ---> thrombin

    (a) 3, 4, 1, 2
    (b) 1, 2, 3, 4
    (c) 4, 3, 1, 2
    (d) 3, 2, 1, 4
    (e) 3, 4, 2, 1

(d)   33. Which of the following secretions is NOT properly matched with its source?
    (a) liver, albumin
    (b) kidney, erythropoietin
    (c) lymphocyte, antibody
    (d) platelet, fibrinogen
    (e) stomach, intrinsic factor

(b)   34. Cells that ingest foreign cells or molecules include ALL BUT:

   (a) eosinophils.
   (b) lymphocytes.
   (c) neutrophils.
   (d) monocytes.

35. True/False. Determine whether each statement is true or false. If it is true, put a T in the answer blank. If it is false, correct the underlined word or phrase and write your correction in the answer blank.

(prothrombin activator)   Prothrombin is activated to thrombin by the activity of <u>VII complex</u>.

(platelets or thrombocytes)   Cell fragments that function in hemostasis are <u>megakaryocytes</u>.

(O)   The most common ABO blood group is group <u>AB</u>.

(calcium)   The ion absolutely essential for the clotting process is <u>sodium</u>.

(recipient)   The universal ABO group blood <u>donor</u> is group AB.

(differential)   The best test to determine the relative proportion of WBCs is the <u>total</u> WBC count.

(increase)   A slowed blood flow and roughened blood vessel walls <u>decrease</u> the chance that a clot will be formed.

(fat-soluble)   Vitamin K is a <u>water-soluble</u> vitamin.

(T)   The major danger of the hemolysis that occurs during a transfusion reaction is <u>renal shutdown</u>.

(T)   Bleeding due to thrombocytopenia is evidenced by the appearance of <u>petechiae</u> on the skin.

(increases)   Hypoxia <u>decreases</u> erythropoiesis by stimulating the kidneys to release REF.

(anemia)   Common signs of <u>polycythemia</u> are fatigue and pallor.

(pernicious)   Underproduction of intrinsic factor causes <u>iron-deficiency</u> anemia.

(T)   The common stem cell of all formed elements is the <u>hemocytoblast</u>.

(liver)   Physiologic jaundice reflects immaturity of the newborn's <u>kidneys</u>.

11 The Circulatory System

Answer the following multiple-choice questions by circling the letter of the correct response. For matching questions, write the correct response in the answer blank.

(b) 1. The layer of the heart wall composed of cardiac muscle is the:
  (a) endocardium.
  (b) myocardium.
  (c) epicardium.
  (d) parietal pericardium.

(a) 2. Which of the following is NOT related to the right heart?
  (a) bicuspid valve
  (b) coronary sinus
  (c) pulmonary valve
  (d) SA node

(e) 3. Which of the following statements describes the position of the heart?
  (a) The apex is in the fifth intercostal space.
  (b) The heart is medial to the lungs.
  (c) The base lies beneath the second rib.
  (d) The heart rests on the diaphragm.
  (e) All of the above.

(b) 4. In the fetus, the left atrium receives most of its blood through the:
  (a) ductus arteriosus.
  (b) foramen ovale.
  (c) fossa ovalis.
  (d) pulmonary artery.

(d) 5. The tricuspid valve is closed:
  (a) when the ventricle is in diastole.
  (b) by movement of blood from the atrium to the ventricle.
  (c) while the atrium is contracting.
  (d) when the ventricle is in systole.

(b) 6. The aortic semilunar valve opens when:
  (a) the ventricle is in diastole.
  (b) the arterial pressure is lower than the ventricular pressure.
  (c) the atrium is in systole.
  (d) all of the above.

(a) 7. The proper term for deficiency of blood is:
  (a) ischemia.
  (b) angina pectoris.
  (c) myocardial infarct.
  (d) pericarditis.

(c) 8. The visceral pericardium is also known as the:

    (a) fibrous pericardium.

    (b) endocardium.

    (c) epicardium.

    (d) myocardium.

(b) 9. Stenosis of the mitral valve may cause blood to back up into the:

    (a) venae cavae.

    (b) pulmonary circulation.

    (c) left ventricle.

    (d) coronary circulation.

(b) 10. The second sound of the heart ("dup") occurs when:

    (a) the AV valves close.

    (b) the semilunar valves close.

    (c) blood is ejected from the ventricles.

    (d) blood is ejected from the atria.

(c) 11. Factors that aid venous return include ALL BUT:

    (a) activity of skeletal muscles.

    (b) pressure changes in the thorax.

    (c) greater urinary output.

    (d) venous valves.

(a) 12. Histologically, this tunic is squamous epithelium underlain by a sparse connective tissue layer:

    (a) tunica intima.

    (b) tunica media.

    (c) tunica externa.

    (d) tunica adventitia.

(c) 13. The arteries that directly feed the capillary beds are called:

    (a) muscular arteries.

    (b) elastic arteries.

    (c) arterioles.

    (d) none of the above.

(b) 14. Any damage to the _____ is referred to as heart block.

    (a) SA node

    (b) AV node

    (c) AV bundle

    (d) Purkinje fibers

    (e) AV valves

(d)   15. Hypertension is a problem because:
      (a)   it can contribute to arteriosclerosis.
      (b)   it can cause heart disease and decompensation.
      (c)   it can damage blood vessels and precipitate thrombus formation.
      (d)   all of the above.

(e)   16. The circumflex artery is a branch of the:
      (a)   right coronary artery.
      (b)   celiac trunk.
      (c)   pulmonary trunk.
      (d)   common carotid artery.
      (e)   left coronary artery.

(a)   17. A late sign of shock is:
      (a)   rapidly falling blood pressure.
      (b)   increased heart rate.
      (c)   rapid thready pulse.
      (d)   cold, clammy skin.

(e)   18. Which of the following is likely during vigorous exercise?
      (a)   Capillaries of the active muscles will be engorged with blood.
      (b)   The skin will be warm.
      (c)   Blood will be diverted from the digestive organs.
      (d)   Blood flow to the heart will increase.
      (e)   All of the above.

(b)   19. All arteries of the systemic circulation branch from the:
      (a)   superior vena cava.
      (b)   aorta.
      (c)   pulmonary artery.
      (d)   coronary artery.

(c)   20. The arterial system that supplies the brain with blood is the:
      (a)   hepatic portal system.
      (b)   pulmonary system.
      (c)   carotid system (includes the Circle of Willis).
      (d)   hepatic system.

(d)   21. An obstruction in the inferior vena cava would hamper the return of blood from the:
      (a)   head and neck.
      (b)   upper limbs.
      (c)   thorax.
      (d)   abdomen and pelvis.

(c)   22. Which statement best describes arteries?
      (a)   All carry oxygenated blood to the heart.
      (b)   All contain valves to prevent the backflow of blood.
      (c)   All carry blood away from the heart.
      (d)   Only large arteries are lined with endothelium.

(c) 23. Which of the following are involved in pulmonary circulation?
    (a) superior vena cava, right atrium, left ventricle
    (b) inferior vena cava, right atrium, left ventricle
    (c) right ventricle, pulmonary artery, left atrium
    (d) left ventricle, aorta, inferior vena cava

(a) 24. Permitting the exchange of nutrients and gases between the blood and tissue cells is the primary function of:
    (a) capillaries.
    (b) arteries.
    (c) veins.
    (d) arterioles.

(c) 25. The circulatory route that runs from the digestive tract to the liver is called:
    (a) coronary circulation.
    (b) pulmonary circulation.
    (c) hepatic portal circulation.
    (d) cerebral circulation.

(c) 26. Given the general statement: the larger the cross-sectional area of a segment of a cardiovascular system, the slower the blood flows, the blood flow in capillaries is:
    (a) very fast.
    (b) moderately fast.
    (c) relatively slow.

(a) 27. The P wave of a normal electrocardiogram indicates:
    (a) atrial depolarization.
    (b) ventricular depolarization.
    (c) atrial repolarization.
    (d) ventricular repolarization.

(e) 28. After several hours of standing at attention on a hot sunny day, several army recruits fainted. Compared to normal resting values, measurements taken on the recruits just before they collapsed would have indicated:
    (a) decreased cardiac output.
    (b) increased blood volume in the legs.
    (c) decreased cerebral blood flow.
    (d) decreased venous return.
    (e) all of the above.

(c) 29. The portion of the intrinsic conduction system located in the interventricular septum is the:
    (a) AV node.
    (b) SA node.
    (c) bundle of His.
    (d) Purkinje fibers.

(b) 30. An ECG provides information about:
    (a) cardiac output.
    (b) movement of the excitation wave across the heart.
    (c) coronary circulation.
    (d) valve impairment.

(b) 31. In heart failure, venous return is slowed. Edema results because:
    (a) blood volume is increased.
    (b) venous pressure is increased.
    (c) osmotic pressure is increased.
    (d) none of the above.

(b) 32. Freshly oxygenated blood is received by the:
    (a) right atrium.
    (b) left atrium.
    (c) right ventricle.
    (d) left ventricle.

(c) 33. The left ventricular wall of the heart is thicker than the right wall in order to:
    (a) accommodate a greater volume of blood.
    (b) expand the thoracic cage during diastole.
    (c) pump blood against greater resistance.
    (d) pump blood through a smaller valve.

(b) 34. The chordae tendineae:
    (a) close the atrioventricular valves.
    (b) prevent the AV valve flaps from everting.
    (c) contract the papillary muscles.
    (d) open the semilunar valves.

(d) 35. Which statement does NOT accurately describe veins?
    (a) They have less elastic tissue and smooth muscle than arteries.
    (b) They contain more fibrous tissue than arteries.
    (c) Most veins in the extremities have valves.
    (d) They always carry deoxygenated blood.

(b) 36. Which of the following tissues is mainly responsible for vasoconstriction?
    (a) elastic tissue
    (b) smooth muscle
    (c) collagenic tissue
    (d) adipose tissue

(c) 37. Which of the following can lead to decreased venous return of blood to the heart?
    (a) an increase in blood volume
    (b) an increase in venous pressure
    (c) damage to the venous valves
    (d) increased muscular activity

(e)   38. Which of the following can contribute to increased arterial blood pressure?

        (a)   increasing stroke volume
        (b)   increasing heart rate
        (c)   arteriosclerosis
        (d)   increasing blood volume
        (e)   all of the above

(d)   39. Exchange of nutrients and wastes in the capillary beds is regulated by local chemical controls. Which of the following would result in the opening of the precapillary sphincters to the capillary bed?

        (a)   a decrease in oxygen content of the blood
        (b)   an increase in carbon dioxide content of the blood
        (c)   a local increase in histamine
        (d)   all of the above

(b)   40. The pressoreceptors in the carotid and aortic bodies are sensitive to:

        (a)   a decrease in carbon dioxide.
        (b)   changes in arterial pressure.
        (c)   a decrease in oxygen.
        (d)   all of the above.

(b)   41. The myocardium receives its blood supply directly from:

        (a)   the aorta.
        (b)   the coronary arteries.
        (c)   the coronary sinus.
        (d)   the pulmonary arteries.

(a)   42. Blood flow in the circulation is steady despite the rhythmic pumping action of the heart because of:

        (a)   the elasticity of the large arteries.
        (b)   the small diameter of the capillaries.
        (c)   the thin walls of the veins.
        (d)   the venous valves.

(b)   43. Tracing the blood from the heart to the right hand, we find that blood leaves the heart and passes through the ascending aorta, the right subclavian artery, the axillary and brachial arteries, and through either the radial or ulnar artery to arrive at the hand. Which artery is missing from this sequence?

        (a)   coronary
        (b)   brachiocephalic
        (c)   cephalic
        (d)   right common carotid

(d)   44. Peripheral resistance in the circulatory system:

        (a)   is inversely related to the diameter of the arterioles.
        (b)   tends to increase if blood viscosity increases.
        (c)   is directly proportional to total length of the blood vessels.
        (d)   all of the above.

(c)   45. Which of the following forces promotes the movement of fluid from interstitial fluid into the blood?
   (a)   capillary hydrostatic pressure
   (b)   mean arterial pressure
   (c)   blood osmotic pressure
   (d)   interstitial fluid osmotic pressure

(c)   46. Which of the following blood pressure readings would be indicative of hypertension?
   (a)   120/80
   (b)   140/90 in a 60-year-old woman
   (c)   170/96
   (d)   110/60

(d)   47. Fenestrated capillaries are found in organs specialized for:
   (a)   protection.
   (b)   support.
   (c)   movement.
   (d)   filtration.

(d)   48. The functions of the lymphatic system include:
   (a)   transport of excess tissue fluid to the blood vascular system.
   (b)   transport of "leaked" proteins to the blood vascular system.
   (c)   maintenance of fluid balance in the tissue spaces.
   (d)   all of the above.

(b)   49. The largest lymphoid organ is:
   (a)   a lymph node.
   (b)   the spleen.
   (c)   the thymus.
   (d)   a Peyer's patch.

(a)   50. The largest collections of lymph nodes are in the:
   (a)   cervical, inguinal, and axillary regions.
   (b)   groin.
   (c)   abdomen.
   (d)   cervical and axillary regions.

(b)   51. Which lymphatic system structure receives lymph from the entire left side of the body and from the region of the right side of the body inferior to the diaphragm?
   (a)   right lymphatic duct
   (b)   thoracic duct
   (c)   cisterna chyli
   (d)   lumbar trunk

(a)   52. Which of the following is NOT a normal component of lymph?

   (a) blood cells
   (b) water
   (c) ions
   (d) plasma proteins

53. Using I for increased and D for decreased, indicate whether the heart rate would be increased or decreased by the action of each of the following factors:

   (I) Epinephrine
   (D) Digitalis
   (I) Thyroxine
   (D) Action of vagus nerve
   (I) Decreased venous return
   (I) Action of the sympathetic nervous system
   (I) Increased temperature
   (I) Increase in CO_2

54. Using I for increased and D for decreased, indicate whether the following factors or events will increase or decrease blood pressure (as a rule):

   (I) A high salt diet
   (D) A blow to the head that damages (disables) the vasomotor center
   (I) An attack by a mugger
   (D) A hypothalamic tumor resulting in excessive ADH production

55. Rank the blood vessels listed below according to their relative blood pressure. Use 1 to indicate the highest pressure.

   (1) Aorta
   (5) Femoral vein
   (2) Brachial artery
   (3) Arterioles in muscle tissue
   (6) Superior vena cava
   (4) Capillaries in muscle tissue

56. True/False. Determine whether each statement is true or false. If it is true, put a T in the answer blank. If it is false, correct the underlined word or phrase and write your correction in the answer blank.

   (BP cuff or sphyg-   The instrument used to occlude an artery when
   momanometer)   obtaining a blood pressure reading is a stethoscope.
   (cardiac output)   Stroke volume is the amount of blood pumped out by a
                ventricle in one minute.
   (T)   The single most important factor determining peripheral resistance is the diameter of the blood vessels (especially the arterioles).
   (T)   The thymus decreases in size with age.
   (T)   Lymphatic capillaries are more permeable than blood capillaries.
   (spleen)   The thymus is the so-called red blood cell graveyard.
   (veins)   Lymphatic vessels derive from arteries of the blood vascular system.
   (T)   Only lymph nodes have afferent lymphatics.

(bradycardia) A heart rate less than 60 beats per minute is called <u>tachycardia</u>.

(T) Chest pain with ischemia of heart muscle that is temporary is called <u>angina</u>.

(T) Generally speaking, only <u>sympathetic</u> fibers are important in regulating the diameter of the blood vessels.

(decrease) An <u>increase</u> in the peripheral resistance will decrease arterial blood pressure.

(medulla) The vasomotor center is located in the <u>pons</u> of the brain stem.

(less) The hydrostatic pressure at the venous end of the capillary bed is <u>greater</u> than it is at the arterial end of the capillary bed.

(T) Generation of angiotensin II from a blood protein follows the release of <u>renin</u> by the kidneys.

(T) Norepinephrine and acetylcholine exert <u>cooperative</u> effects on the heart.

(T) Cardiac output does not decrease during periods of sympathetic nervous system control because both heart rate and <u>length</u> of contraction are increased by norepinephrine.

(thinner) The walls of veins are <u>thicker</u> than the walls of corresponding arteries.

(SA node) The heart tissue with the fastest intrinsic rate of depolarization is the <u>AV node</u>.

(T) Propagation of the depolarization wave through the <u>AV node</u> is delayed for approximately 0.1 sec to allow the atria to finish their contraction.

12   Body Defenses

Answer the following multiple-choice questions by circling the letter of the correct response. For matching questions, write the correct response in the answer blank.

(c)   1. All of the following are considered nonspecific body defenses EXCEPT:

    (a) complement.

    (b) phagocytosis.

    (c) antibodies.

    (d) lysozyme.

    (e) inflammation.

(d)   2. Antibodies released by plasma cells are involved in:

    (a) humoral immunity.

    (b) immediate hypersensitive reactions.

    (c) autoimmune disorders.

    (d) all of the above.

(c)   3. Complement fixation:

    (a) lyses the cell to which it is attached.

    (b) enhances phagocytosis.

    (c) enhances inflammation.

    (d) all of the above.

(a)   4. Which of the following is not an effect of histamine and kinin?

    (a) vasoconstriction

    (b) increased capillary permeability

    (c) positive chemotaxis

    (d) activation of pain receptors

(b)   5. The process whereby neutrophils and other white blood cells are attracted to an inflammatory site is called:

    (a) diapedesis.

    (b) chemotaxis.

    (c) margination.

    (d) phagocytosis.

(e)   6. Antibodies are characterized by all of the following EXCEPT:

    (a) protein molecules.

    (b) antigen-specific.

    (c) composed of heavy and light polypeptide chains.

    (d) two or more binding sites per molecule.

    (e) incapable of being transferred from one person to another.

(c) 7. All of the following are typical of complete antigens EXCEPT:
   (a) large molecules.
   (b) chemically complex.
   (c) contain many repeating chemical units.
   (d) stimulate production of antibodies.
   (e) react with the product of the immune response.

(d) 8. Which of the following factors/events does NOT enhance or amplify the normal activity of the immune response?
   (a) complement
   (b) monokines
   (c) activity of neutrophils
   (d) suppressor factors

(d) 9. When antigens interact with IgE antibodies attached to mast cells:
   (a) degranulation of the mast cell occurs.
   (b) local capillaries become leaky.
   (c) a hypersensitivity response occurs.
   (d) all of the above.

(d, e) 10. Which of the following antibodies can fix complement?
   (a) IgA
   (b) IgD
   (c) IgE
   (d) IgG
   (e) IgM

(a) 11. Which antibody class is abundant in body secretions?
   (a) IgA
   (b) IgD
   (c) IgE
   (d) IgG
   (e) IgM

(c) 12. The B cell response to the initial antigen challenge includes all of the following EXCEPT:
   (a) enlargement and rapid mitosis.
   (b) formation of a clone.
   (c) immediate production of antigen-specific antibodies.
   (d) production of progeny cells that include plasma cells and memory cells.

(d) 13. Autoimmune disorders may result from:
   (a) exposure to previously "hidden" self-antigens to the immunocompetent cells.
   (b) cross-reaction of antibodies formed against foreign antigens with self antigens.
   (c) mutation followed by the appearance of membrane proteins not previously present.
   (d) all of the above.

(a)   14. Passive immunity involves all EXCEPT:

    (a)   long term immune protection.

    (b)   short-term immune protection.

    (c)   infusion of immune serum.

    (d)   passage of IgG antibodies from a pregnant mother to her fetus.

(b)   15. An injection of attenuated pathogen would produce immunity that is:

    (a)   active and natural.

    (b)   active and artificial.

    (c)   passive and natural.

    (d)   passive and artificial.

(b)   16. The process by which antibodies link soluble antigens into large lattices which then become insoluble and settle out of solution is called:

    (a)   agglutination.

    (b)   precipitation.

    (c)   neutralization.

    (d)   complement fixation.

(e)   17. Types of T cells include:

    (a)   cytotoxic.

    (b)   memory.

    (c)   helper.

    (d)   suppressor.

    (e)   all of the above.

(d)   18. The immune response is involved with all EXCEPT:

    (a)   destruction of cancer cells that arise within the body.

    (b)   rejection of foreign tissue grafts.

    (c)   disposal of cells housing viruses or other intracellular parasites.

    (d)   stockpiling toxic chemicals.

(a)   19. Small molecules that bind with self-proteins to produce antigenic substances are called:

    (a)   haptens.

    (b)   antibodies.

    (c)   ions.

    (d)   reagins.

(b)   20. Lymphocytes that develop immunocompetence in the thymus are:

    (a)   B lymphocytes.

    (b)   T lymphocytes.

(c)   21. Cells that can directly lyse target cells include all of the following EXCEPT:

    (a)   macrophages.

    (b)   cytotoxic T cells.

    (c)   helper T cells.

    (d)   natural killer cells.

(d) 22. Which of the following is involved in the activation of a B cell?

   (a) antigen
   (b) helper T cell
   (c) lymphokines
   (d) all of the above

(c) 23. One of Alice's kidneys was transplanted into her son. This type of graft is a(n):

   (a) autograft.
   (b) isograft.
   (c) allograft.
   (d) zenograft.

24. True/False. Determine whether each statement is true or false. If it is true, put a T in the answer blank. If it is false, correct the underlined word or phrase and write your correction in the answer blank.

(plasma cells) The effector cells that produce the bulk of the antibodies released are <u>B cells</u>.

(IgM) The pentamer immunoglobulin is <u>IgA</u>.

(variable) The <u>constant</u> regions of the immunoglobulin chains are the antigen-binding regions.

(T) The <u>constant regions</u> of the antibody structure determine immunoglobulin class.

(T) A person who is infected with a particular bacterium and suffers the disease develops/displays <u>active</u> immunity.

(macrophages) Accessory cells that act as antigen presenters thus helping to activate the lymphocytes are <u>neutrophils</u>.

(T) The protein molecules that protect tissue cells from viral "takeover" are collectively called <u>interferons</u>.

(T cells) Lymphokines are produced by <u>macrophages</u>.

(80) Approximately <u>50</u> percent of blood-borne immunoglobulins are IgG.

(suppressor) The T cell responsible for tempering and shutting down the immune response when the pathogen has been routed is the <u>helper</u> T cell.

(autoimmune disease) The failure of the immune system to tolerate self results in <u>immunodeficiencies</u>.

(edema or swelling) The four cardinal signs of inflammation are heat, redness, pain, and <u>immobility</u>.

(T) A group of cells all descended from the same ancestral cells is called a <u>clone</u>.

(T) The initial response to antigen is called the <u>primary</u> response.

(immunocompetent) When antigen-specific receptors first appear on developing lymphocytes, they are said to have become <u>reactive</u>.

13   The Respiratory System

Respond to the following multiple-choice questions by circling the letter of the correct response. For matching questions, write the correct response in the answer blank.

1. Match the terms in Column B with the descriptions in Column A.

Column A	Column B
(d) Flushing of air into and out of the lungs.	(a) respiratory gas transport
(c) Exchange of respiratory gases that occurs between systemic capillaries and tissue cells.	(b) external respiration
	(c) internal respiration
	(d) pulmonary ventilation
(a) Movement of oxygen and carbon dioxide to and from the lungs and tissue cells.	
(b) Exchange of respiratory gases that occurs in the lungs.	

(d) 2. Which of the following is (are) a function(s) of the mucus-covered nasal passages?

(a) warming inspired air
(b) moistening inspired air
(c) entrapping small particles
(d) all of the above

(b) 3. The lung volume that represents the total volume of exchangeable air is the:

(a) tidal volume.
(b) vital capacity.
(c) inspiratory capacity.
(d) expiratory reserve volume.
(e) residual air.

(a) 4. The amount of air that enters or leaves the lungs during a normal respiratory cycle is the:

(a) tidal volume.
(b) residual volume.
(c) vital capacity.

(e) 5. Possible causes of hypoxia include:

(a) too little oxygen in the atmosphere.
(b) obstruction of the trachea.
(c) pneumonia.
(d) a large amount of functional "dead space."
(e) all of the above.

(c)   6. The pressure within the lungs is called:
  (a) intra-abdominal pressure.
  (b) intraplural pressure.
  (c) intrapulmonary pressure.
  (d) intrathoracic pressure.

(c)   7. All of the following are portions of the respiratory zone EXCEPT:
  (a) alveolar sacs.
  (b) alveolar ducts.
  (c) terminal bronchioles.
  (d) respiratory bronchioles.

(b)   8. All of the following are true about the control of respiration by oxygen EXCEPT:
  (a) mediated by chemoreceptors.
  (b) sensitive enough that normal changes in alveolar ventilation have profound effects on the amount of oxygen carried in the blood.
  (c) relatively weak compared to the effect of carbon dioxide.
  (d) becomes very important in cases of COPD, e.g., emphysema and chronic bronchitis.

(c)   9. The normal pH of the blood is about:
  (a) 8.4.
  (b) 7.8.
  (c) 7.4.
  (d) 4.7.

(b)   10. At the structural level, the event that promotes inspiration is:
  (a) the increase in intrapulmonary pressure.
  (b) the decrease in intrapulmonary pressure.
  (c) the increase in intrapleural pressure.
  (d) all of the above.

(b)   11. Death from CO (carbon monoxide) poisoning would be due to:
  (a) the body's destruction of Hb.
  (b) the strong combination of CO on the same Hb sites with which O_2 normally combines.
  (c) formation of cellular toxins.
  (d) inability to form hemoglobin.

(b)   12. The walls of the alveoli are composed of:
  (a) ciliated columnar epithelium.
  (b) simple squamous epithelium.
  (c) stratified squamous epithelium.
  (d) loose connective tissue.

(e)   13. Examples of nonrespiratory air movements include:

       (a) coughs.

       (b) yawns.

       (c) sneezes.

       (d) hiccups.

       (e) all of the above.

(a)   14. The epiglottis:

       (a) prevents aspiration of food into the lower respiratory passages.

       (b) is located in the nasopharynx.

       (c) is composed of hyaline cartilage.

       (d) is active in both alert and comatose people.

(d)   15. Which of the following is used by the respiratory system and the digestive system?

       (a) esophagus

       (b) larynx

       (c) trachea

       (d) pharynx

(c)   16. The larynx is "roofed" by the:

       (a) cricoid cartilage.

       (b) thyroid cartilage.

       (c) epiglottis.

       (d) costal cartilage.

(a)   17. The Adam's apple is a common term for the:

       (a) thyroid cartilage.

       (b) cricoid cartilage.

       (c) epiglottis.

       (d) none of the above.

(d)   18. The C-shaped rings of cartilage of the trachea not only prevent the trachea from collapsing but also aid in the process of:

       (a) lubrication.

       (b) removing foreign particles.

       (c) gas exchange.

       (d) swallowing.

(d)   19. The respiratory membrane consists of:

       (a) a single thickness of epithelial cells.

       (b) a single thickness of epithelial cells and a basement membrane.

       (c) two thicknesses of epithelial cells.

       (d) two thicknesses of epithelial cells and their fused basement membranes.

(b)   20. Cutting the phrenic nerves will result in:

       (a) air entering the pleural cavity.

       (b) paralysis of the diaphragm.

       (c) stimulation of the diaphragmatic reflex.

       (d) paralysis of the epiglottis.

(a)   21. Following the removal of his larynx, an individual would:
      (a) be unable to speak.
      (b) be unable to cough.
      (c) have difficulty swallowing.
      (d) be in respiratory difficulty or arrest.

(c)   22. Under ordinary circumstances, the Hering-Breuer reflex is initiated by:
      (a) the inspiratory center.
      (b) the apneustic center.
      (c) overinflation of the alveoli and bronchioles.
      (d) the pneumotaxic center.

(c)   23. The detergentlike molecule that keeps the alveoli from collapsing between breaths because it reduces the surface tension of the water film in the alveoli is called:
      (a) lecithin.
      (b) bile.
      (c) surfactant.
      (d) reluctant.

(d)   24. When the inspiratory muscles contract:
      (a) the size of the thoracic cavity is increased in diameter.
      (b) the size of the thoracic cavity is increased in length.
      (c) the volume of the thoracic cavity is decreased.
      (d) the size of the thoracic cavity is increased in both length and diameter.

(b)   25. Oxygen and carbon dioxide are exchanged in the lungs and through all cell membranes by:
      (a) active transport.
      (b) diffusion.
      (c) filtration.
      (d) osmosis.

(c, d) 26. Which of the following would NOT normally be treated by 100% oxygen therapy? (Choose all that apply.)
      (a) anoxia
      (b) carbon monoxide poisoning
      (c) respiratory crisis in an emphysema patient
      (d) eupnea

(d)   27. Carbonic anhydrase is:
      (a) a CO_2 carrier molecule found in plasma.
      (b) an enzyme that degrades carbon monoxide (CO).
      (c) a molecule that reduces alveolar surface tension.
      (d) an enzyme in RBCs that converts CO_2 to bicarbonate.

(c)   28. Most oxygen carried in the blood is:
    (a) in solution in the plasma.
    (b) combined with plasma proteins.
    (c) chemically combined with the heme in red blood cells.
    (d) in solution in the red blood cells.

(b)   29. Which of the following has the greatest stimulating effect on the respiratory center in the brain?
    (a) oxygen
    (b) carbon dioxide
    (c) calcium
    (d) willpower

(a)   30. In mouth-to-mouth artificial respiration, the rescuer blows air from his or her own respiratory system into that of the victim. Which of the following statements are correct?
    1. Expansion of the victim's lungs is brought about by blowing air in at higher than atmospheric pressure (positive-pressure breathing).
    2. During inflation of the lungs, the intrapleural pressure increases.
    3. This technique will not work if the victim has a hole in the chest wall, even if the lungs are intact.
    4. Expiration during this procedure depends on the elasticity of the lungs.

    (a) all of the above
    (b) 1, 2, 4
    (c) 1, 2, 3
    (d) 2, 4

(b)   31. A baby holding its breath will:
    (a) have brain cells damaged because of low blood oxygen levels.
    (b) automatically start to breathe again when the carbon dioxide levels in the blood reach a high enough value.
    (c) suffer heart damage because of increased pressure in the carotid sinus and aortic arch areas.
    (d) be termed a "blue baby."

(c)   32. Under ordinary circumstances, which of the following blood components is of NO physiological significance?
    (a) bicarbonate ions
    (b) carbaminohemoglobin
    (c) nitrogen
    (d) chloride

(b)   33. Damage to which of the following would result in cessation of breathing?
    (a) the pneumotaxic center
    (b) the medulla
    (c) the stretch receptors in the lungs
    (d) the apneustic center

(b)   34. The bulk of carbon dioxide is carried:

    (a)   chemically combined with the amino acids of hemoglobin as carbamino hemoglobin in the red blood cells.

    (b)   as the ion HCO_3 in the plasma after first entering the red blood cell.

    (c)   as carbonic acid in the plasma.

    (d)   chemically combined with the heme portion of Hb.

(b)   35. Halted or difficult breathing is termed:

    (a)   apnea

    (b)   dyspnea

    (c)   hyperpnea

    (d)   eupnea

(c)   36. Carbon dioxide:

    1.   concentration in the blood is decreased by voluntary hyperventilation.

    2.   accumulation in the blood is associated with an increase in acidity.

    3.   will not stimulate ventilation if the nerves from the peripheral (and aortic body) chemoreceptors have been cut.

    4.   concentration is greater in the venous blood than in the arterial blood.

    (a)   all of the above

    (b)   1, 2, 3

    (c)   1, 2, 4

    (d)   2, 3, 4

(d)   37. Cystic fibrosis is associated with:

    (a)   underactive respiratory control centers.

    (b)   muscle spasms in bronchial walls.

    (c)   lack of alveolar surfactant.

    (d)   production of thick mucus which clogs small respiratory and digestive passages.

38. True/False. Determine whether each statement is true or false. If it is true, put a T in the answer blank. If it is false, correct the underlined word or phrase and write your correction in the answer blank.

    (lower) The intrapleural pressure is always <u>higher</u> than the intrapulmonary pressure.

    (T) Inspiration occurs when the intrapulmonary pressure becomes <u>lower</u> than the atmospheric pressure.

    (blue) Cyanosis causes a <u>cherry-red</u> skin color due to inadequate oxygen delivery to the tissues.

    (inhibits) The pneumotaxic center generally <u>stimulates</u> inspiration.

    (chronic bronchitis) In <u>emphysema</u>, the important pathology is inflammation of the respiratory passages and their obstruction with mucus.

    (alveoli) The actual site of gas exchange is the <u>bronchioles</u>.

    (smoking) The major factor contributing to lung cancer and COPD is <u>diet</u>.

14   The Digestive System and Body Metabolism

Answer the following multiple-choice questions by circling the letter of the correct response. For matching questions, write the correct response in the answer blank.

(d)   1. Which of the following is NOT true of the duodenum?

        (a)   It receives chyme from the stomach.
        (b)   It is the site of action of secretions of the liver and pancreas.
        (c)   It is shorter than the ileum or jejunum.
        (d)   It is an important site of mechanical digestion.

(b)   2. The correct sequence for the layers within the wall of the alimentary canal from inside to outside is:

        (a)   mucosa, muscularis, serosa, submucosa.
        (b)   mucosa, submucosa, muscularis, serosa.
        (c)   serosa, muscularis, mucosa, submucosa.
        (d)   submucosa, mucosa, serosa, muscularis.

(c)   3. Which of the following is NOT a function of the mouth?

        (a)   mastication
        (b)   secretion of amylase
        (c)   absorption of glucose
        (d)   deglutition

(a)   4. Which of the following statements is FALSE?

        (a)   The substance surrounding the pulp cavity is enamel.
        (b)   The root of the tooth is embedded in the alveolus of the jaw.
        (c)   Blood vessels and nerves are contained within the central pulp.
        (d)   A periodontal membrane secures the tooth to bone.

(d)   5. The esophagus:

        (a)   extends from the pharynx to the stomach.
        (b)   is approximately 10 inches long in the adult.
        (c)   is posterior to the trachea, anterior to the vertebral column, and pierces the diaphragm.
        (d)   all of the above.

(b)   6. Which class of foods significantly delays stomach emptying?

        (a)   proteins
        (b)   fats
        (c)   carbohydrates
        (d)   fluids

(d)   7. Surgical cutting of the frenulum would affect which part of the body?

       (a)   salivary glands
       (b)   esophagus
       (c)   nasal cavity
       (d)   tongue

(a)   8. The portion of the large intestine closest to the liver is the:

       (a)   ascending colon.
       (b)   rectum.
       (c)   transverse colon.
       (d)   descending colon.

(c)   9. Which of the following is NOT characteristic of the large intestine?

       (a)   It is divided into ascending, transverse, and descending portions.
       (b)   It contains abundant bacteria, some of which synthesize certain vitamins.
       (c)   It is the main digestive site.
       (d)   It absorbs much of the water and salts remaining in the wastes.

(c)   10. The gallbladder:

       (a)   produces bile.
       (b)   is attached to the pancreas.
       (c)   stores and concentrates bile.
       (d)   produces secretin.

(a)   11. The sphincter between the stomach and duodenum is:

       (a)   the pyloric sphincter.
       (b)   the cardiac sphincter.
       (c)   the sphincter of Oddi.
       (d)   the ileocecal sphincter.

(c)   12. Chemical digestion of carbohydrates is initiated in the:

       (a)   stomach.
       (b)   small intestine.
       (c)   mouth.
       (d)   large intestine.

(b)   13. The main chemical activity of the stomach is to begin the digestion of:

       (a)   fats.
       (b)   proteins.
       (c)   carbohydrates.
       (d)   all of the above.

(d)   14. In the absorptive phase of digestion, a comparison of blood entering and leaving the liver shows that:

       (a)   blood leaving the liver contains more glucose.
       (b)   blood leaving the liver contains more amino acids.
       (c)   the liver has little effect on the nutrient composition of the blood.
       (d)   blood leaving the liver usually has fewer ingested toxic substances.

(a)   15. Which of the following events occurs while swallowing?

        (a) peristalsis
        (b) closure of the gastroesophageal sphincter
        (c) segmentation
        (d) opening of the glottis

(c)   16. Chemical digestion reduces large complex molecules to simpler compounds by the process of:

        (a) oxidation-reduction reactions.
        (b) anabolism.
        (c) hydrolysis.
        (d) fermentation.

(a)   17. The site of production of gastrin is:

        (a) the stomach.
        (b) the small intestine.
        (c) the pancreas.
        (d) the large intestine.

(b)   18. Carbohydrates are acted on by:

        (a) peptidases, trypsin, and chymotrypsin.
        (b) amylase, maltase, and sucrase.
        (c) lipases.
        (d) peptidases, lipases, and galactose.

(b)   19. The parasympathetic nervous system influences digestion by:

        (a) relaxing smooth muscle.
        (b) stimulating peristalsis and secretory activity.
        (c) constricting sphincters.
        (d) none of the above.

(a)   20. The digestive juice product containing enzymes capable of digesting all three major foodstuff categories is:

        (a) pancreatic.
        (b) gastric.
        (c) salivary.
        (d) biliary.

(d)   21. The vitamin associated with calcium absorption is:

        (a) A.
        (b) K.
        (c) C.
        (d) D.

(d) 22. Someone has eaten a meal of buttered toast, cream, and eggs. Which of the following would you expect to happen?

   (a) Compared to the period shortly after the meal, gastric motility and secretion of HCl decrease when the food reaches the duodenum.
   (b) Gastric motility increases even while the person is chewing the food (before any swallowing).
   (c) Fat will be emulsified in the duodenum by the action of bile.
   (d) All of the above.

(b) 23. The site of production of cholecystokinin is:

   (a) the stomach.
   (b) the small intestine.
   (c) the pancreas.
   (d) the large intestine.

In questions 24–28, trace the path of a single protein molecule that has been ingested.

(d) 24. The protein molecule will be digested by enzymes secreted by:

   (a) the mouth, stomach, and colon.
   (b) the stomach, liver, and small intestine.
   (c) the small intestine, mouth, and liver.
   (d) the pancreas, small intestine, and stomach.

(d) 25. The protein molecule must be digested before it can be transported to and utilized by the cells because:

   (a) protein is only useful directly.
   (b) protein has a low pH.
   (c) proteins in the circulating blood produce an adverse osmotic pressure.
   (d) the protein is too large to be readily absorbed.

(b) 26. The products of protein digestion enter the bloodstream largely through cells lining:

   (a) the stomach.
   (b) the small intestine.
   (c) the large intestine.
   (d) the bile duct.

(c) 27. Before the blood carrying the products of protein digestion reaches the heart, it first passes through capillary networks in:

   (a) the spleen.
   (b) the lungs.
   (c) the liver.
   (d) the brain.

(a)   28. Having passed through the regulatory organ selected above, the products of protein digestion are circulated through the body. They will enter individual body cells as a result of:

    (a)   active transport.

    (b)   diffusion.

    (c)   osmosis.

    (d)   pinocytosis.

(d)   29. The secretion of saliva is stimulated by:

    (a)   chewing.

    (b)   the presence of food (chemicals) in the mouth.

    (c)   the mental picture or odor of a liked food.

    (d)   all of the above.

(b)   30. A fluid present in the small intestine during digestion that contains cholesterol, bilirubin, and phospholipids is:

    (a)   pancreatic juice.

    (b)   bile.

    (c)   intestinal juice.

    (d)   gastric juice.

(b)   31. The layer of the digestive tube that contains blood vessels, lymphatic nodes, and glands is the:

    (a)   mucosa.

    (b)   submucosa.

    (c)   muscularis externa.

    (d)   serosa.

(b)   32. As the food in the stomach is repeatedly squeezed and mixed with gastric juice, it is converted into a semifluid consistency called:

    (a)   a bolus.

    (b)   chyme.

    (c)   bile.

    (d)   feces.

(d)   33. Those structures in the small intestine that increase its surface area and thus aid in the absorption of food are called:

    (a)   villi.

    (b)   microvilli.

    (c)   plicae circulares.

    (d)   all of the above.

(a)   34. The acidic contents entering the small intestine are buffered by:

    (a) HCO_3 in pancreatic juice.

    (b)   bile.

    (c)   intestinal juice.

    (d)   pancreatic enzymes.

(b) 35. Which of the following is NOT an illustration of digestion?

    (a) polysaccharides --> disaccharides

    (b) carbon dioxide and water --> carbohydrates

    (c) fat --> fatty acids and glycerol

(c) 36. The end-products of the hydrolysis of carbohydrates are:

    (a) amino acids.

    (b) fatty acids.

    (c) monosaccharides.

    (d) proteoses.

(b) 37. What is the relationship, if any, between the process of digestion and that of cellular respiration?

    (a) Respiration provides the organic molecules used in digestion.

    (b) Digestion provides the simpler molecules required for respiration.

    (c) None, because digestion refers to food.

    (d) Waste products of respiration leave the body via the digestive tract.

(c) 38. The formation of glucose from glycogen is:

    (a) gluconeogenesis.

    (b) glycogenesis.

    (c) glycogenolysis.

    (d) glycolysis.

(d) 39. Which of the following best defines cellular respiration?

    (a) intake of carbon dioxide and output of oxygen by cells

    (b) excretion of waste products

    (c) inhalation of oxygen and exhalation of carbon dioxide

    (d) oxidation of substances by which energy is released in usable form to the cells

(c) 40. Metabolic rate is relatively low in:

    (a) youth.

    (b) physical exercise.

    (c) old age.

    (d) fever.

(a) 41. In a temperate climate under ordinary conditions, the greatest loss of body heat occurs through:

    (a) radiation.

    (b) conduction.

    (c) evaporation.

    (d) none of the above.

(d) 42. Which of the following is NOT a function of the liver?

    (a) glycogenolysis and gluconeogenesis

    (b) synthesis of cholesterol

    (c) detoxification of alcohol and drugs

    (d) synthesis of glucagon

    (e) deamination of amino acids

(a)   43. Defecation is triggered by:
    (a)   stretching of the wall of the large intestine.
    (b)   activity of the vagus nerve.
    (c)   secretion of cholecystokinin.
    (d)   the enterogastric reflex.

(c)   44. Functions of cholesterol include all of the following EXCEPT:
    (a)   component of plasma membrane.
    (b)   vitamin D synthesis.
    (c)   source of energy.
    (d)   steroid hormone synthesis.

(d)   45. Amino acids are essential (and important) to the body for all the following EXCEPT:
    (a)   production of some hormones.
    (b)   production of antibodies.
    (c)   formation of most structural materials.
    (d)   as a source of quick energy.

(a)   46. A person has been on a hunger strike for 7 days. Compared to normal, he has:
    (a)   increased release of fatty acids from adipose tissues and ketosis.
    (b)   elevated glucose concentration in the blood.
    (c)   increased plasma insulin concentration.

(d)   47. Three days after removal of the pancreas from an animal, the researcher finds a persistent increase in:
    (a)   ketone concentration in the blood.
    (b)   urine volume.
    (c)   blood glucose.
    (d)   all of the above.

(b)   48. Vitamins:
    (a)   are organic compounds that serve as building blocks for structural components and are sometimes used for energy.
    (b)   function as coenzymes to assist in catalysis.
    (c)   are also called provitamins if used in catalytic reactions.
    (d)   all of the above.

49. True/False. Determine whether each statement is true or false. If it is true, put a T in the answer blank. If it is false, correct the underlined word or phrase and write your correction in the answer blank.

    (T)   For the most part, we depend on microorganisms in the large intestine to synthesize vitamin K for us.

    (constipation) Too much water absorption by the large intestine will result in diarrhea.

    (mesentery) The double-layered folds of the peritoneum that supports portions of the small intestine are called the lesser omentum.

    (T)   The process that causes the breakdown of fat globules into smaller droplets is emulsification.

(pepsinogen) HCl in the stomach is used to activate the gastric enzyme precursor called amylase.

(hepatic or common bile) Jaundice is one symptom of obstruction of the pancreatic duct.

(absorption) The transfer of digested foodstuffs from the lumen of the alimentary canal to the blood is called active transport.

(32) Permanent teeth number 36.

(vomiting) Reverse peristalsis, which moves ingested food toward the mouth, is called segmentation.

(enzymes) Biological catalysts involved in chemical digestion are called substrates.

(parotid) Mumps is an inflammation of the submandibular salivary glands.

(T) The fang-like teeth adapted for piercing or tearing food are the canines.

(Periodontal disease) Dental caries account for 80 to 90% of tooth loss in adults.

(parasympathetic) The major neural controls of digestion are exerted by the sympathetic division of the autonomic nervous system.

(mucus) Ulcers are a common consequence when hydrochloric acid production by the stomach is inadequate.

(microvilli) The brush border enzymes of the small intestine are located in the crypts of Lieberkuhn.

(cholecystokinin or CCK) The hormone that prods the pancrease to release its juice and stimulates the gall bladder to contract is secretin.

(metabolism) The term that means all chemical reactions necessary to maintain life is catabolism.

(T) The lipoproteins that are "good" because they transport cholesterol from the blood vessel walls (or tissue cells) to the liver for disposal are the high-density lipoproteins.

(T) Eventually all of the energy derived from foodstuffs is converted to heat.

(basal metabolic rate, or BMR) The amount of energy used by the body to maintain vital activities only is called the total metabolic rate, or TMR.

(thyroxine) The single most important hormonal influence on metabolic rate is epinephrine.

(hypothalamus) The body's "thermostat" for maintaining homeostatic body temperature is located in the medulla oblongata.

(rising) Cool, pale skin and shivering indicate that body temperature is falling.

(appendicitis) The major inflammatory condition affecting adolescents is gastritis.

(fats) Of the three groups of foods—carbohydrates, fats, and proteins—carbohydrates yield the most calories per gram.

(seven; about a dozen) The body requires adequate supplies of only three minerals and trace amounts of all others.

15 The Urinary System

Answer the following multiple-choice questions by circling the letter of the correct response. For matching questions, write the correct response in the answer blank.

(b) 1. Urine passes through the:

    (a) kidney hilus to bladder to ureter.

    (b) renal pelvis to ureter to bladder to urethra.

    (c) glomerulus to ureter to renal tubule.

    (d) hilus to urethra to bladder.

(a) 2. The outermost portion of the kidney is the:

    (a) cortex.

    (b) pelvis.

    (c) medulla.

    (d) pyramids.

(a) 3. The functional unit of the kidney is the:

    (a) nephron.

    (b) ureter.

    (c) urethra.

    (d) hilus.

(a) 4. In the glomerular capsule, one afferent arteriole divides into a capillary network called a(n):

    (a) glomerulus.

    (b) interlobular artery.

    (c) arcuate artery.

    (d) efferent arteriole.

(b) 5. An obstruction in the glomerulus would affect the flow of blood into the:

    (a) renal artery.

    (b) efferent arteriole.

    (c) afferent arteriole.

    (d) interlobular artery.

6. Match the terms in Column B with the descriptions in Column A.

Column A	Column B
(b) Site of filtrate formation.	(a) collecting tubule
(a) Structure that conveys the processed filtrate to the renal pelvis.	(b) glomerulus
(c) peritubular capillaries
(m) Its inner membrane forms part of the filtration membrane.	(d) distal convoluted tubule
(e) proximal convoluted tubule
(e) Primary site of tubular reabsorption.	(f) interlobar artery
(g) interlobular artery
(d) Secondarily important site of tubule reabsorption.	(h) arcuate artery
(i) lobar artery
(c) Blood that directly receives substances from the tubular cells.	(j) efferent arteriole
(k) afferent arteriole
(h) Receives blood from the interlobar artery.	(l) loop of Henle
(m) glomerular capsule
(i) A direct branch of the renal artery.

(b) 7. The epithelium of the urinary bladder that permits distension is:

    (a) stratified squamous.

    (b) transitional.

    (c) simple squamous.

    (d) pseudostratified columnar.

(d) 8. Which of the following regions of the nephron may be found in the renal medulla?

    (a) glomerulus

    (b) proximal convoluted tubule

    (c) distal convoluted tubule

    (d) loop of Henle

(c) 9. The filtration membrane includes ALL BUT:

    (a) glomerular endothelium.

    (b) podocytes.

    (c) renal fascia.

    (d) basement membrane.

(d) 10. Which of the following is NOT considered a function of the kidneys?

    (a) the regulation of body fluid concentration

    (b) the regulation of body fluid volume

    (c) the removal of waste products from the body

    (d) the formation of plasma proteins

(c)   11. The fluid in the glomerular capsule (Bowman's capsule) is similar to plasma except that it has little:

    (a) glucose.
    (b) hormones.
    (c) protein.
    (d) electrolytes.

(c)   12. If you remove the posterior lobe of the pituitary from an experimental animal, a copious flow of urine develops. How would you explain this effect on the kidneys?

    (a) The hypothalamus contains a center for the control of water balance; possibly, you have injured the hypothalamus also.
    (b) You may have cut the nerves to the kidneys.
    (c) You have removed the source of ADH release.
    (d) The blood aids in regulating water balance; possibly, you have increased the blood flow through the kidneys.

(c)   13. The kidneys are stimulated to produce renin:

    (a) when the peritubular capillaries are dilated.
    (b) when the pH of the urine decreases.
    (c) by a decrease in the blood supply to the kidneys.
    (d) when the specific gravity of urine rises above 1.10.

(a)   14. The function of angiotensin is to:

    (a) constrict arterioles.
    (b) decrease the production of aldosterone.
    (c) decrease arterial blood pressure.
    (d) decrease water absorption.

(d)   15. Select the correct statements from the following:

    1. Most of the water passing through the kidneys is eliminated as urine.
    2. Normally filtrate contains a large amount of protein.
    3. Some resorption of the water is hormonally controlled.
    4. The secretion of hydrogen ions is one of the mechanisms that maintains the pH of the blood.

    (a) 1, 2, 4
    (b) 2, 3, 4
    (c) 1, 4
    (d) 3, 4

(b)   16. Filtration at the glomerulus is directly related to:

    (a) water reabsorption.
    (b) arterial blood pressure.
    (c) capsular hydrostatic pressure.
    (d) acidity of the urine.

(e)   17. Tubular secretions adds which of the following substance(s) to the filtrate?

       (a) glucose

       (b) H^+

       (c) K^+

       (d) water

       (e) both b and c

(c)   18. If a freshly voided urine sample contains excessive amounts of urochrome, it has:

       (a) an ammonialike odor.

       (b) a pH below normal.

       (c) a dark yellow color.

       (d) a pH above normal.

(a)   19. Conditions such as diabetes mellitus, starvation, and low-carbohydrate diets are closely linked to:

       (a) ketosis.

       (b) pyuria.

       (c) albuminuria.

       (d) hematuria.

(d)   20. The lowest blood concentration of nitrogenous waste occurs in the:

       (a) hepatic vein.

       (b) inferior vena cava.

       (c) renal artery.

       (d) renal vein.

(b)   21. The capillaries of the glomerulus differ from other capillary networks in the body because they:

       (a) have a larger area of anastomosis.

       (b) are derived from and drain into arterioles.

       (c) are not made of endothelium.

       (d) cause the filtrate to be forced from the blood.

(c)   22. Damage to the renal medulla would interfere first with the functioning of the:

       (a) glomerular capsules.

       (b) distal convoluted tubules.

       (c) collecting tubules.

       (d) proximal convoluted tubules.

(d)   23. Which is reabsorbed by the proximal convoluted tubule?

       (a) Na^+

       (b) K^+

       (c) amino acids

       (d) all of the above

(c)   24. Glucose is not normally found in the urine because it:
   (a)   does not pass through the walls of the glomerulus.
   (b)   is kept in the blood by colloid osmotic pressure.
   (c)   is reabsorbed by the tubule cells.
   (d)   is removed by the body cells before the blood reaches the kidney.

(b)   25. An abnormal constituent of urine is:
   (a)   urea.
   (b)   hemoglobin.
   (c)   ammonia.
   (d)   creatinine.

(b)   26. Physiological acidosis might be caused by:
   (a)   hyperventilation.
   (b)   renal excretion of H^+.
   (c)   renal excretion of bicarbonate ions ($CHCO_3^-$).
   (d)   both a and c.

(d)   27. Which of the following depend on the presence of electrolytes?
   (a)   maintenance of osmotic relations between cells and ECF
   (b)   control of H^+ concentrations by buffer action
   (c)   enzyme activity
   (d)   all of the above

(a)   28. Aldosterone is secreted in responses to:
   (a)   low ECF sodium.
   (b)   high ECF volumes.
   (c)   low ECF potassium.
   (d)   all of the above.

(a)   29. Aldosterone causes:
   (a)   increased renal tubular resorption of sodium.
   (b)   decreased renal secretion of potassium.
   (c)   decreased resorption of chloride.
   (d)   increased permeability of the distal convoluted tubule to water.

(a)   30. To relax after the final examination, a student goes to a local bar and drinks two
          quarts of beer (!!!) in which there is a minimal amount of sodium. Which of the fol-
          lowing changes will occur?
   (a)   a large increase in plasma volume
   (b)   a decrease in solute concentration of body fluids and in secretion of ADH
   (c)   a decrease in aldosterone secretion by the adrenal cortex
   (d)   an increase in the osmolarity of the urine

(a)   31. Body water content is greatest in:

      (a) infants.
      (b) young adults.
      (c) elderly adults.

(c)   32. Potassium, magnesium, and phosphate ions are the predominant electrolytes in:

      (a) plasma.
      (b) interstitial fluid.
      (c) intracellular fluid.

(b)   33. Sodium balance is regulated primarily by control of the amount(s):

      (a) ingested.
      (b) excreted in urine.
      (c) lost in perspiration.
      (d) lost in feces.

(b)   34. Water balance is regulated primarily by control of the amount(s):

      (a) ingested.
      (b) excreted in urine.
      (c) lost in perspiration.
      (d) lost in feces.

(a)   35. The single most important electrolyte involved in determining fluid shifts and distribution of body water is:

      (a) sodium.
      (b) potassium.
      (c) bicarbonate.
      (d) chloride.

(c)   36. The fluid link between the external and internal environment is:

      (a) interstitial fluid.
      (b) intracellular fluid.
      (c) blood plasma.
      (d) cerebrospinal fluid.
      (e) lymph.

(b)   37. Which of the following is NOT a major chemical buffer system in the body?

      (a) bicarbonate
      (b) renal
      (c) phosphate
      (d) protein buffer system

(d)   38. Which of the following statements is false?

      (a) The external urethral sphincter is made of skeletal muscle.
      (b) The internal urethral sphincter is made of smooth muscle.
      (c) The bladder wall contains smooth muscle.
      (d) The control center for micturition is located in the medulla oblongata.

39. True/False. Determine whether each statement is true or false. If it is true, put a T in the answer blank. If it is false, correct the underlined word or phrase and write your correction in the answer blank.

(H+/hydrogen ion) Chemical buffers act to prevent dramatic changes in <u>carbon dioxide</u> concentration.

(T) The lungs along with the kidneys can be recognized as <u>excretory</u> organs.

(T) The kidneys help maintain the internal <u>chemical balance</u> of the body.

(posterior lobe) The hormone ADH is released from the <u>anterior lobe</u> of the pituitary and helps in the reabsorption of water from the collecting ducts.

(nucleic acids) Uric acid represents one of the nitrogenous waste products resulting from the metabolism of <u>proteins</u>.

(retroperitoneal) The kidneys are positioned <u>within the peritoneal cavity</u> at the level of the twelfth pair of ribs.

(T) Kinking of the ureter is a common consequence of an abnormal descent of the kidney, or <u>ptosis</u>.

(proximal convoluted tubules) The cells of the <u>collecting tubules</u> have dense microvilli which reveal their important role in reabsorption.

(low) An important function characteristic of the peritubular capillaries is their <u>high</u> hydrostatic pressure.

(nephron) The <u>renal corpuscle</u> is the functional unit of the kidney.

(juxtaglomerular) The source of renin is the <u>juxtamedullary</u> apparatus of the nephrons.

(smaller) Efferent arterioles have a <u>larger</u> diameter than afferent arterioles.

(stimulates) Sympathetic nervous system stimulation of the kidneys <u>inhibits</u> renin release.

(Pyuria) <u>Proteinuria</u> indicates an infection somewhere in the urinary tract.

(anuria) Urinary output of less than 100 mL/day is called <u>oliguria</u>.

(1.5 L to 1.8 L) Although about 180 L of filtrate are formed daily by the kidneys, only about <u>2.5 L</u> of urine are excreted in the same time period.

(diabetes insipidus) The condition of excessive urinary output due to a deficit of ADH is called <u>diabetes mellitus</u>.

(external) The <u>internal</u> urethral sphincter formed of skeletal muscle is a voluntary sphincter.

(bladder) Micturition means emptying the <u>bowel</u>.

16　The Reproductive System

Answer the following multiple-choice questions by circling the letter of the correct response. For matching questions, write the correct response in the answer blank.

(d)　　1. The scrotum is:

    (a)　divided medially.

    (b)　an adaptation to provide for temperature homeostasis of the testes.

    (c)　a homologue of the labia majora of females.

    (d)　all of the above.

(c)　　2. The interstitial cells of the testes:

    (a)　produce sperm.

    (b)　dispose of spermatid excess cytoplasm.

    (c)　produce testosterone.

    (d)　all of the above.

(e)　　3. In the human male:

    (a)　sperm are stored in the epididymis.

    (b)　spermatogenesis is the division process that forms spermatids.

    (c)　a vasectomy is performed to terminate the release of sperm.

    (d)　following castration, the accessory reproductive organs decrease in size.

    (e)　all of the above.

(d)　　4. All of the following statements about spermatogenesis are true EXCEPT:

    (a)　the spermatogonium forms the primary spermatocyte.

    (b)　the primary spermatocyte forms two secondary spermatocytes.

    (c)　the secondary spermatocytes each form two spermatids.

    (d)　each spermatid forms two sperm.

(d)　　5. All of the following statements about sperm are true EXCEPT:

    (a)　formation results from both meiotic and mitotic division of the male germinal epithelium.

    (b)　may remain viable (after ejaculation) for up to 72 hours.

    (c)　formation is inhibited at normal body temperature.

    (d)　are inactivated by an alkaline environment.

(a)　　6. A low secretion of luteinizing hormone (LH) in the normal male adult would cause:

    (a)　inhibition of testosterone secretion.

    (b)　excessive beard growth.

    (c)　increased spermatogenesis.

    (d)　shrinkage of the anterior pituitary gland.

7. Match the key terms with the descriptions that follow.

Key: (a) ductus deferens   (c) seminiferous tubules of the testes
      (b) epididymis   (d) urethra

(c) Site of sperm production.

(b) Site where sperm mature and become motile.

(a) Upon ejaculation, sperm are propelled through the sperm duct or
_____.

(d) The pathway to the body exterior is completed by the _____.

(b) 8. The process by which a spermatid is converted to a motile sperm is called:

  (a) spermatogenesis.
  (b) spermiogenesis.
  (c) meiosis.
  (d) mitosis.

(c) 9. Erection of the penis results from:

  (a) contraction of muscle.
  (b) gubernaculum action.
  (c) vascular dilation in the erectile tissue of the penis.
  (d) willpower.

(d) 10. Which will occur if the testes do not descend?

  (a) Male sex hormones will not be circulated in the body.
  (b) Sperm will have no means of exit from the body.
  (c) Inadequate blood supply will retard the development of the testes.
  (d) Viable sperm will not be produced.

(c) 11. The normal number of human chromosomes is:

  (a) 48.
  (b) 47.
  (c) 46.
  (d) 23.
  (e) 24.

(c) 12. Spermatozoa are to seminiferous tubules as oocytes are to:

  (a) fimbriae.
  (b) corpus albicans.
  (c) ovarian follicles.
  (d) corpora lutea.

(a) 13. The acrosome:

  (a) releases enzymes that help a sperm penetrate an egg.
  (b) induces peristalsis of male ducts.
  (c) is made of microtubules.
  (d) contains DNA.

(d)   14. Which of the following does NOT add a secretion that contributes to semen?

   (a) prostate
   (b) bulbourethral glands
   (c) testes
   (d) vas deferens

(a)   15. Female gonads are the:

   (a) ovaries.
   (b) uterine tube.
   (c) uterus.
   (d) vagina.

(b)   16. The corpus luteum is formed at the site of:

   (a) fertilization.
   (b) ovulation.
   (c) menstruation.
   (d) implantation.

(a)   17. The sex of a child is determined by:

   (a) the sex chromosome contained in the sperm.
   (b) the sex chromosome contained in the oocyte.
   (c) the number of sperm fertilizing the oocyte.
   (d) the position of the fetus in the uterus.

(a)   18. The menstrual cycle can be divided into three continuous phases. Starting from the first day of the cycle, their consecutive order is:

   (a) menstrual, proliferative, secretory.
   (b) menstrual, secretory, proliferative.
   (c) secretory, menstrual, proliferative.
   (d) proliferative, menstrual, secretory.

(b)   19. The structures that draw an ovulated oocyte into the female duct system are:

   (a) cilia.
   (b) fimbriae.
   (c) microvilli.
   (d) stereocilia.

(a)   20. The usual site of embryo implantation is:

   (a) the uterine tube.
   (b) the peritoneal cavity.
   (c) the vagina.
   (d) the uterus.

(a)   21. The male structure that corresponds to the female clitoris is:

   (a) the penis.
   (b) the scrotum.
   (c) the penile urethra.
   (d) the testis.

(a) 22. Which of the following is NOT included in the vulva?

    (a) fundus
    (b) labia
    (c) clitoris
    (d) greater vestibular glands

(d) 23. Which of the following is correct relative to female anatomy?

    (a) The vaginal orifice is the most dorsal of the three openings in the vulva.
    (b) The urethra is between the vaginal orifice and the anus.
    (c) The anus is between the vaginal orifice and the urethra.
    (d) The urethra is the most ventral of the three orifices.

(b) 24. Secondary sex characteristics are:

    (a) present in the embryo.
    (b) a result of male or female sex hormones increasing in amount at puberty.
    (c) the testis in the male and the ovary in the female.
    (d) not subject to withdrawal once established.

(d) 25. Estrogens are produced by:

    (a) the placenta.
    (b) the ovary.
    (c) the pharmaceutical companies.
    (d) all of the above.

(c) 26. The compartments of the scrotum communicate with the abdominopelvic cavity through the:

    (a) excretory ducts.
    (b) testes.
    (c) inguinal canals.
    (d) penis.

(b) 27. The soft mucosal lining of the uterus is the:

    (a) peritoneum.
    (b) endometrium.
    (c) perimetrium.
    (d) myometrium.

(d) 28. An interesting hormonal control mechanism involves the corpus luteum. This structure normally degenerates toward the end of the monthly cycle, and menstruation follows. However, if the ovum is fertilized and starts to develop, the corpus luteum remains active temporarily, and, of course, there is no menstruation. What hormone exerts this influence on the corpus luteum?

    (a) estrogen
    (b) progesterone
    (c) LH
    (d) human chorionic gonadotropin (HCG)

(e) 29. Testosterone stimulates:

    (a) growth of muscle and the skeleton.

    (b) facial hair growth.

    (c) growth of the larynx.

    (d) enlargement of the male genitalia and accessory structures.

    (e) all of the above.

(e) 30. Effects of estrogen include:

    (a) enlargement of the vagina and labia.

    (b) female distribution of fatty tissue.

    (c) growth of the breasts.

    (d) growth of pubic hair.

    (e) all of the above.

(c) 31. Secretion of progesterone stimulates:

    (a) secretory activity of the glands of the breasts.

    (b) contraction of uterine muscles.

    (c) secretory activity of the glands of the uterine endometrium.

    (d) development of the female secondary sex characteristics.

(b) 32. A boy who has not passed through puberty sustains an injury to his anterior pituitary such that FSH is no longer released, but LH secretion is normal. After he grows to maturity, one would expect that he would:

    (a) not develop secondary sex characteristics.

    (b) be sterile.

    (c) be impotent (unable to have an erection).

    (d) all of the above.

(c) 33. The most important hormone in initiating and maintaining lactation after birth is:

    (a) estrogen.

    (b) FSH.

    (c) prolactin.

    (d) oxytocin.

(d) 34. Human ova and sperm are similar in that:

    (a) about the same number of each is produced per month.

    (b) they have the same degree of motility.

    (c) they are about the same size.

    (d) they have the same number of chromosomes.

(b) 35. All of the following are true about the human female process of reproduction EXCEPT:

    (a) the monthly discharge of the uterine lining (menses) is initiated by the fall in secretion of female hormones.

    (b) ovulation always occurs 15 days after the beginning of menses.

    (c) an overactive adrenal cortex may cause growth of facial hair in the female.

    (d) the uterine tubes are open to the peritoneal cavity.

(d)　36. Which of the following occur after ovulation?
　　　　(a)　The corpus luteum secretes estrogen and progesterone.
　　　　(b)　The endometrium enters its secretory phase.
　　　　(c)　Secretion of anterior pituitary gonadotropins is inhibited.
　　　　(d)　All of the above.

(a)　37. The primary reproductive organs are:
　　　　(a)　the testes and ovaries.
　　　　(b)　also called the external genitalia.
　　　　(c)　the testes and penis in males.
　　　　(d)　the breasts and vulva in females.

(a)　38. The sex of a new individual is determined:
　　　　(a)　at the moment of fertilization.
　　　　(b)　at birth.
　　　　(c)　by the fetal stage.
　　　　(d)　at different times depending on the genetic makeup of the individual.

(b)　39. A test to detect cancerous changes in cells of the uterus and cervix is a:
　　　　(a)　pyelogram.
　　　　(b)　pap smear.
　　　　(c)　D & C.
　　　　(d)　laparoscopy.

(d)　40. All of the following are true about ovulation EXCEPT:
　　　　(a)　results in the discharge of the oocyte from the ovary into the abdominal cavity.
　　　　(b)　is triggered by a sudden increase in LH secretion.
　　　　(c)　is blocked during pregnancy.
　　　　(d)　will not occur in the absence of progesterone produced by the corpus luteum.

(a)　41. All of the following statements referring to the menstrual cycle are true EXCEPT:
　　　　(a)　FSH and LH directly promote development of the uterine endometrium.
　　　　(b)　Estrogen is secreted by the developing follicle in the first phase of the cycle.
　　　　(c)　The corpus luteum is formed from the ruptured follicle after ovulation.
　　　　(d)　The emotional state of the female may markedly alter the output of FSH and LH.

(b)　42. Sex chromosomes of a normal male are:
　　　　(a)　XX.
　　　　(b)　XY.
　　　　(c)　YY.
　　　　(d)　any of these depending on the father.

(b)　43. A zygote is usually formed within the:
　　　　(a)　ovary.
　　　　(b)　uterine (fallopian) tube.
　　　　(c)　uterus.
　　　　(d)　endometrium.

(d)   44. The trophoblast and the inner cell mass are layers of the:
  (a)   ovary.
  (b)   uterus.
  (c)   placenta.
  (d)   blastocyst.

(c)   45. The embryonic stage that is a hollow "ball of cells" is the:
  (a)   zygote.
  (b)   morula.
  (c)   blastocyst.
  (d)   fetus.

(d)   46. The embryonic stage lasts:
  (a)   2 weeks.
  (b)   4 weeks.
  (c)   6 weeks.
  (d)   8 weeks.

(c)   47. Cleavage as part of embryonic development is distinctive because it involves:
  (a)   the fusion of gametes.
  (b)   meiosis.
  (c)   cell division by mitosis with little or no growth between successive divisions.
  (d)   all of the above.

(b)   48. Implantation involves:
  (a)   fusion of egg and sperm.
  (b)   embedding of the young embryo in the uterine wall.
  (c)   formation of germ layers.
  (d)   placenta completion.

(b)   49. The embryo is directly enclosed in and protected by:
  (a)   the yolk sac.
  (b)   the amnion.
  (c)   the allantois.
  (d)   the chorion.

(a)   50. The umbilical arteries carry:
  (a)   waste products to the placenta.
  (b)   oxygen and food to the fetus.
  (c)   oxygen and food to the placenta.
  (d)   oxygen and waste products to the fetus.

(b)   51. In the fetal circulation, blood bypasses the nonaerated lungs via the:
  (a)   ductus venosus.
  (b)   foramen ovale.
  (c)   vena cava.
  (d)   umbilical vein.

(e) 52. Maternal changes during pregnancy include:
   (a) weight gain.
   (b) increased metabolic rate.
   (c) increased blood volume.
   (d) increased cardiac output.
   (e) all of the above.

(b) 53. Nutrition of the young embryo before implantation occurs is primarily provided by:
   (a) nutrients stored in the embryonic cells.
   (b) simple diffusion of nutrients from the uterine secretions.
   (c) the blood supply of the placenta.
   (d) all of the above.

(c) 54. During pregnancy, the corpus luteum normally produces estrogens and progesterone:
   (a) only until the implantation of the embryo.
   (b) until the eighth week of pregnancy.
   (c) for about 8 to 12 weeks.
   (d) for the entire pregnancy.

(d) 55. The onset of labor may be a result of all the following factors EXCEPT:
   (a) stretch of uterine and cervical muscles by the baby.
   (b) decline in progesterone production by the deteriorating placenta.
   (c) secretion of oxytocin.
   (d) increased estrogen production.
   (e) increased prostaglandin activity.

(d) 56. All of the following are true about events preceding fertilization EXCEPT:
   (a) enzymes (acrosomal) of the sperm disperse the cells of the corona radiata surrounding the oocyte.
   (b) the head of the sperm forms the male pronucleus.
   (c) the secondary oocyte completes the second meiotic division.
   (d) the two pronuclei divide before fusing.

57. Using the key below, indicate from which germ layer each of the named structures or organs develops.
   Key: (a) mesoderm   (b) endoderm   (c) ectoderm

   (a) Bony skeleton
   (a) Circulatory system
   (c) Nervous system
   (b) Lining of the GI, respiratory, and reproductive tracts
   (a) Skeletal muscles

(c) 58. The first stage of labor is to dilation of the cervix as the second stage is to:
   (a) expulsion of the placenta.
   (b) escape of the amniotic fluid.
   (c) delivery of the infant.
   (d) uterine contractions.

(b) 59. The copulatory organ of females is the:

    (a) uterus.
    (b) vagina.
    (c) clitoris.
    (d) all of the above.

(a) 60. The source of nutrition for sperm is fructose in:

    (a) seminal vesicle fluid.
    (b) prostatic fluid.
    (c) the secretion of bulbourethral glands.
    (d) all of these.

(b) 61. "Hot flashes," osteoporosis, and increased risk of heart attack are all consequences of:

    (a) labor
    (b) menopause
    (c) menses
    (d) pregnancy

(e) 62. The placenta is the _____ organ for the fetus:

    (a) excretory.
    (b) nutritive.
    (c) respiratory.
    (d) endocrine.
    (e) all of the above.

63. True/False. Determine whether each statement is true or false. If it is true, put a T in the answer blank. If it is false, correct the underlined word or phrase and write your correction in the answer blank.

(T) The production of <u>testosterone</u> by an embryo results in the formation of the male accessory reproductive organs.

(albuginea) The tunica <u>vaginalis</u> is the tough fibrous connective tissue capsule that surrounds the testis.

(T) During puberty, the production of <u>estrogen</u> by the newly awakened ovaries results in the appearance of the secondary sex characteristics in females.

(alkaline or basic) Semen is <u>acidic</u>.

(bulbourethral glands) Mucus is secreted by the <u>seminal vesicles</u>.

(cryptorchidism) Failure of the testes to descend is called <u>mittelschmerz</u>.

(puberty) The period of life when the reproductive system becomes functional is referred to as <u>menopause</u>.

(T) The Golgi apparatus of a maturing sperm elaborates an enzyme-containing "helmet" over the tip of the nucleus called an <u>acrosome</u>.

(breast) The leading cause of death from cancer in women is <u>cervical</u> cancer.

(cervix) The muscular outlet of the uterus is the <u>fundus</u>.

(vertex) The head-down presentation of a fetus is called the <u>breech</u> presentation.

(T) The first menstrual period in a young woman is called <u>menarche</u>.

(seminal vesicle) The ejaculatory duct is formed by the fusion of the duct of the <u>prostate gland</u> and the end of the ductus deferens.

(circumcision) Surgical removal of the foreskin is called a <u>vasectomy</u>.

(Graafian) The mature ovarian follicle is called a <u>secondary</u> follicle.